T0141212

CITES Bulb Checklist

For the genera:

Cyclamen, *Galanthus* and *Sternbergia*

Prepared and edited by:

Aaron P Davis

and

H Noel McGough, Brian Mathew & Christopher Grey-Wilson

Royal Botanic Gardens, Kew

First published in 1999

General editor of series: Jacqueline A Roberts

ISBN 1 900347 39 3

Compiled with the financial assistance of:

the CITES Nomenclature Committee
the Royal Botanic Gardens, Kew
the Netherlands Ministry of Agriculture, Nature Management and Fisheries

Front cover: *Galanthus ikariae*
(photograph by A P Davis)

Cover design by Media Resources RBG Kew
Printed and bound by Whitstable Litho Ltd, Whitstable, Kent

Acknowledgements

The authors would like to thank: Johan van Scheepen (Koninklijke Algemeene Vereeniging voor Bloembollencultuur—the Royal General Bulbgrowers Association—The Netherlands) for reviewing the checklist; Hayri Duman (Gazi Ün. Fen-Ed. Fak., Ankara, Turkey), Tuna Ekim (Istanbul Üni. Fen Fak., Botanik ABD, Istanbul, Turkey), Neriman Özhatay (Istanbul Üni. Eczacilik Fak., Farmasötik Botanik BD, Istanbul, Turkey), Estrela Figueiredo (Herbário, Centro de Botânico, Instituto de Innestigaçâo Científica Tropical, Lisbon, Portugal), Helen Mordak (Komarov Botanical Institute, St Petersburg, Russia) and Dick Brummitt, Jill Cowley, Rafaël Govaerts, Jacqui Roberts, Suzy Dickerson and Mike Lock (Royal Botanic Gardens, Kew, United Kingdom) for their help during the preparation and publication of the checklist.

We would particularly like to thank The Netherlands Ministry for Agriculture, Nature Management and Fisheries for their financial support of the publication of the checklist—without their help this project would not have taken place.

Remerciements

Les auteurs tiennent à remercier Johan van Scheepen (Koninklijke Algemeene Vereeniging voor Bloembollencultuur—the Royal General Bulbgrowers Association—Pays-Bas), qui a revu et corrigé la liste; Hayri Duman (Gazi Ün. Fen-Ed. Fak., Ankara, Turquie), Tuna Ekim (Istanbul Üni. Fen Fak., Botanik ABD, Istanbul, Turquie), Neriman Özhatay (Istanbul Üni. Eczacilik Fak., Farmasötik Botanik BD, Istanbul, Turquie), Estrela Figueiredo (Herbário, Centro de Botânico, Instituto de Innestigaçâo Científica Tropical, Lisbonne, Portugal), Helen Mordak (Komarov Botanical Institute, Saint-Pétersbourg, Russie) et Dick Brummitt, Jill Cowley, Rafaël Govaerts, Jacqui Roberts, Suzy Dickerson et Mike Lock (Royal Botanic Gardens, Kew, Royaume-Uni) pour leur assistance dans la préparation et la publication de la liste.

Nous adressons des remerciements particuliers au Ministère néerlandais de l'Agriculture, de la Gestion de la nature et de la Pêche pour son appui financier à la publication de la liste. Sans cet appui, le projet n'aurait pas pu être mené à bien.

Agradecimientos

Los autores desean dar las gracias a Johan van Scheepen (Koninklijke Algemeene Vereeniging voor Bloembollencultuur—the Royal General Bulbgrowers Association—Países Bajos) por la revisión de la lista, a Hayri Duman (Gazi Ün. Fen-Ed. Fak., Ankara, Turquía), Tuna Ekim (Istanbul Üni. Fen Fak., Botanik ABD, Istanbul, Turquía), Neriman Özhatay (Istanbul Üni. Eczacilik Fak., Farmasötik Botanik BD, Estambul, Turquía), a Estrela Figueiredo (Herbário, Centro de Botânico, Instituto de Innestigaçâo Científica Tropical, Lisboa, Portugal), a Helen Mordak (Komarov Botanical Institute, San Petesburgo, Rusia) y a Dick Brummitt, Jill Cowley, Rafaël Govaerts, Jacqui Roberts, Suzy Dickerson y Mike Lock (Royal Botanic Gardens, Kew, Reino Unido) por su asistencia en la preparación y publicación de la Lista.

En particular, deseamos dar las gracias al Ministerio de Agricultura, Gestión de la Naturaleza y Pesca de los Países Bajos por el apoyo financiero prestado para publicar esta lista, ya que sin él esta publicación no habría visto la luz.

CONTENTS

Preamble

Contents

        A.P.Davis and R.Govaerts

TABLE DES MATIERES

Table des matieres

                    A.P.Davis and R.Govaerts

INDICE

Preámbulo

Indice

         A.P.Davis and R.Govaerts

CITES CHECKLIST - BULBS

PREAMBLE

1. Background

The 1992 Conference of the Parties to the Convention on International Trade in Endangered Species of Wild Fauna and Flora (CITES) adopted the *CITES Cactaceae Checklist* as the guideline when making reference to the species of the genera concerned. This was the first CITES plant checklist, which was followed by the publication of the *CITES Orchid Checklist Volume 1*, in 1995.

These references have proved to be an important tool in the day to day implementation of CITES for plant species. The combination of support from the CITES Conferences of the Parties, individual party states, scientific institutions and organisations has facilitated the preparation and publication of the *CITES Cactaceae Checklist (second edition)*, the *CITES Orchid Checklist Volume 2*, the *CITES Checklist of succulent Euphorbia taxa*, and the *CITES Bulb Checklist*. All of these works have been adopted by the tenth meeting of the Conference of the Parties, as the guidelines when making reference to the accepted names of the genera concerned.

This checklist is the result of co-operation between the Royal Botanic Gardens, Kew (United Kingdom) and the Netherlands Ministry of Agriculture, Nature Management and Fisheries (The Netherlands). Taking advantage of recent work on the horticulturally important and highly traded 'bulb' groups, Aaron Davis was contracted by RBG Kew to bring this work together and to research some problem areas. The Netherlands Ministry of Agriculture provided financial support for the publication of the checklist.

The *CITES Bulb Checklist* is based upon contemporary revisions for *Cyclamen* (Grey-Wilson, 1988 & 1997), *Galanthus* (Davis, 1999) and a new synopsis for the genus *Sternbergia* (prepared by Mathew and Davis, and presented here). This checklist incorporates the largest collection of published synonyms for these three genera. Further data, revisions, edits, and updates have been added during the compilation of this list. This checklist is therefore up-to-date and should be useful for at least several years.

The word 'bulb' is used here in its everyday sense, that is, to represent plants with rounded, swollen underground parts, but excluding orchids. This is not the strict scientific meaning of the term bulb.

2. Methodology

The checklist was compiled in four stages:
- for each genus, data were extracted from the most recent taxonomic revisions and entered into a database
- the database was expanded by including all known published names for each genus (using *Index Kewensis*), original publications and in some instances herbarium specimens
- draft accounts of each genus were produced and these were reviewed

Cyclamen	Chris Grey-Wilson & Aaron Davis
Galanthus	Aaron Davis
Sternbergia	Brian Mathew & Aaron Davis

- the completed checklist was extracted from the database and prepared for camera-ready copy using *Microsoft Word for Windows version 7©*.

3. How to use the checklist

The main aim of this list is to provide a quick reference for checking accepted names, synonymy and distribution.

It has been necessary to include autonyms in this checklist. Autonyms are names that are created automatically when a subspecific name (below the level of species) is added to a species name. For example: in recognising *G. bortkewitschianus* as a variety of *Galanthus alpinus* (giving, *Galanthus alpinus* var. *bortkewitschianus*), *Galanthus alpinus* has its epithet repeated at the same rank as the subspecific addition (giving, *Galanthus alpinus* var. *alpinus*—which is the autonym). Autonyms are included in Parts I, II and III. In Annex I the autonyms are included in square brackets to avoid confusion.

The reference is divided into three main parts:

Part I: ALL NAMES IN CURRENT USE

An alphabetical list of all accepted names and synonyms for the three genera included in this checklist. In Part I, the author's name appears after each taxon where the taxon name appears twice or more (unless the author's name is the same) e.g., *Cyclamen cyprium* Glasau, and *Cyclamen cyprium* Sibth.

Part II: ACCEPTED NAMES IN CURRENT USE

These are separate lists for each genus. Each list is ordered alphabetically by the accepted name and details are given on current synonyms and distribution. Additional information is given by localising the area of distribution in each country: e.g., N, S, E or W, by listing the main islands, and by giving details on introductions. This information is also included in Part III, with a supplement of data on introductions.

Part III: COUNTRY CHECKLIST

Accepted names for all the genera included in this checklist are ordered alphabetically under country of distribution.

4. Conventions employed in parts I, II and III

a) Accepted names are given in **bold roman** type.
Synonyms are given in *italic* type.

b) Where a synonym occurs twice, but refers to different accepted names, e.g., *Galanthus ikariae* subsp. *latifolius*, for both **Galanthus platyphyllus** and **Galanthus woronowii**, the name with an asterisk is the species most likely to be encountered in trade under this name. For example:

All Names	Accepted Name
Galanthus ikariae subsp. *latifolius*	**Galanthus platyphyllus**
Galanthus ikariae subsp. *latifolius*	**Galanthus woronowii***
Galanthus ikariae subsp. *snogerupii*	**Galanthus ikariae**

*Species most likely to be in trade (in this example, **Galanthus woronowii**).

c) Where an accepted name and a synonym have the same epithet, but refer to different species, e.g., **Cyclamen persicum** and *Cyclamen persicum* (**Cyclamen graecum** subsp. **graecum** forma **graecum**), the name with an asterisk is the species most likely to be found in trade under this name. For example:

All Names	Accepted Name
Cyclamen pentelici	**Cyclamen graecum** subsp. **graecum** forma **graecum**
Cyclamen persicum*	
Cyclamen persicum Sibth. & Sm.	**Cyclamen graecum** subsp. **graecum** forma **graecum**

*Species most likely to be in trade (in this example, **Cyclamen persicum**).

NB: In examples b) and c) it is necessary to double-check by reference to the distribution as detailed in Part II.

d) A selection of hybrids have been included in the checklist and are indicated by the multiplication sign ×. They are arranged alphabetically within Parts I, II and III.

5. Number of names entered for each genus:
Cyclamen (accepted 65, synonyms 179); *Galanthus* (accepted 27, synonyms 225); *Sternbergia* (accepted 9, synonyms 35).

6. Abbreviations of book titles, journals and authors' names
Authors' names are abbreviated according to Brummitt and Powell (1992); journal and periodical titles according to Lawrence *et al.* (1968) and Bridson and Smith (1991) and book titles according to Stafleu and Cowan (1976 to 1988), with some minor alterations.

Brummitt, R.K. and Powell, C.E. (1992). *Authors of Plant Names.* Royal Botanic Gardens, Kew.

Lawrence, G.H.M., Günther Buchheim, A.F., Daniels, G.S. and Dolezal, H. (1968). *B-P-H. Botanico-Periodicum-Huntianum.* Pittsburgh: Hunt Botanical Library.

Bridson, G.D.R. and Smith, E.R. (1991). *B-P-H/S Botanico-Periodicum-Huntianum/ Supplementum.* Pittsburgh: Hunt Institute for Botanical Documentation.

Stafleu, F.A. and Cowan, R.S. (1976 - 1988). *Taxonomic Literature*, edn. 2, volumes 1– 7. Bohn, Scheltema and Holkema: Utrecht/Antwerpen; dr. W. Junk b.v. Publishers: The Hague/Boston.

7. Abbreviations, botanical terms, and Latin

Note: words in *italics* are Latin

ambiguous name a name which has been applied to different taxa by different authors, so that it has become a source of ambiguity

anon. anonymous; without author or author unknown

auct. *auctorum*: of authors

cultivation the raising of plants by horticulture or gardening; not immediately taken from the wild

cultivar an individual, or assemblage of plants maintaining the same distinguishing features, which has been produced or is maintained (propagated) in cultivation

descr. *descriptio*: the description of a species or other taxonomic unit

distribution where plants are found (geographical)

ed. editor

edn. edition (book or journal)

eds. editors

epithet the last word of a species, subspecies, or variety (etc.), for example: *lutea* is the species epithet for the species *Sternbergia lutea* and *byzantinus* is the subspecific epithet for *Galanthus plicatus* subsp. *byzantinus*

escape a plant which has left the boundaries of cultivation (e.g. a garden) and is found occurring in natural vegetation

ex *ex*: after; may be used between the names of two authors, the second of whom validly published the name indicated or suggested by the first

excl. *exclusus*: excluded

hort. *hortorum*: of gardens (horticulture); raised or found in gardens; not a plant of the wild

in prep. in preparation

in sched. *in scheda*: on a herbarium specimen or label

in syn. *in synonymia*: in synonymy

incl. including

ined. *ineditus*: unpublished

introduction a plant which occurs in a country, or any other locality, due to human influence (by purpose or chance); any plant which is not native

key a written system used for the identification of organisms (e.g. plants)

leg. *legit*: he gathered; the collector

misspelling a name that has been incorrectly spelt; not a new or different name

morphology the form and structure of an organism (e.g. a plant)

name causing confusion a name that is not used because it cannot be assigned unambiguously to a particular taxon (e.g. a species of plant)

native an organism (e.g. a plant) that occurs naturally in a country, or region, etc.

naturalized a plant which has either been introduced (see introduction) or has escaped (see escape) but which looks like a wild plant and is capable of reproduction in its new environment

nom. *nomen*: name

nom. ambig. *nomen ambiguum*: ambiguous name

nom. cons. prop. *nomen conservandum propositum*: name proposed for conservation under the rules of the International Code for Botanical Nomenclature (ICBN)

nomenclature branch of science concerned with the naming of organisms (e.g. plants)

non *non*: not

only known from cultivation a plant which does not occur in the wild, only in cultivation

orthographic variant an alternative spelling for the same name

pro parte *pro parte*: partly, in part

provisional name name given in anticipation of a valid description

sens. *sensu*: in the sense of; the manner in which an author interpreted or used a name
sens. lat. *sensu lato*: in the broad sense; a taxon (usually a species) and all its subordinate taxa (e.g. subspecies) and/or other taxa sometimes considered as distinct
sic *sic*, used after a word which looks wrong or absurd, to show that it has been quoted correctly
synonym a name that is applied to a taxon but which cannot be used because it is not the accepted name—the synonym or synonyms form the synonymy
taxa plural of taxon
taxon a named unit of classification, e.g. genus, species, subspecies

8. Geographical areas
Country names follow the United Nations' standard as laid down in Country Names. *Terminology Bulletin* August 1995. United Nations 347:1–41.

9. Annex I–VI
There are six annexes: (I) list of all names, with author and place of publication; (II) keys to species; (III) bibliography; (IV) list of extra synonyms for *Galanthus nivalis*; (V) summary of accepted names; (VI) a new combination in *Cyclamen*.

Annex I is a list of all the names and their original place of publication, i.e., where and when they were legitimately named, and by whom. This is intended for those requiring more information on these ornamental plants.

The keys in Annex II provide a means for the identification of *Cyclamen*, *Galanthus* (including subspecies and varieties) and *Sternbergia.*

Annex III is a bibliography of the primary references used in this checklist: this gives access to other information not included here, such as ecology, cultivation details, and descriptions of morphology.

Annex IV provides a list of the more obscure synonyms of *Galanthus nivalis*, and specifically for those names (including many cultivars) originating in the horticultural literature.

Annex V provides an enumeration of all accepted names, enabling the user to see at a glance what is contained within each of the genera.

Annex VI gives the necessary information and details for a new combination in *Cyclamen.*

LISTE DES PLANTES A BULBES CITES

PRÉAMBULE

1. Historique
En 1992, la Conférence des Parties à la Convention sur le commerce international des espèces de faune et de flore sauvages menacées d'extinction (CITES) a adopté la *CITES Cactaceae Checklist* (liste des Cactacées CITES) en tant que référence aux espèces de cette famille. C'était une première liste de plantes couvertes par la CITES, suivie, en 1995, de la *CITES Orchid Checklist* (liste des Orchidées CITES), *Volume 1*.

Ces références se sont révélées très utiles pour la mise en oeuvre pratique de la Convention. L'appui combiné de la Conférence des Parties à la CITES, de Parties individuelles, de certaines institutions et d'organisations scientifiques (IOS, par exemple) a facilité la préparation et la publication de la *CITES Cactaceae Checklist (deuxième édition)*, de la *CITES Orchid Checklist Volume 2*, de la *CITES Checklist of Succulent Euphorbia Taxa* et de la *CITES Bulb Checklist*. Toutes ont été adoptés par la 10ᵉ session de la Conférence des Parties en tant que listes normalisées de référence pour les noms acceptés des taxons concernés.

La présente liste résulte de la coopération des *Royal Botanic Gardens, Kew*, Royaume-Uni (Jardins botaniques royaux, JBR) et du Ministère néerlandais de l'Agriculture, Gestion de la nature et Pêche. Profitant de récents travaux sur ces groupes de plantes à bulbes—travaux importants tant sur le plan horticole que commercial—Aaron Davis a été chargé par les JBR de faire la synthèse de ces travaux et d'approfondir quelques sujets posant des problèmes; le Ministère néerlandais de l'Agriculture a apporté son appui financier à la publication de cette liste.

La *CITES Bulb Checklist* est fondée sur des révisions récentes de *Cyclamen* (Grey-Wilson, 1988 & 1997), *Galanthus* (Davis, 1999) et sur une nouvelle synthèse sur le genre *Sternbergia* (préparée par B.Mathew et A.P.Davis et présentée pour la première fois dans la présente liste).

Le mot "bulbe" est utilisé ici dans son sens général, à savoir pour décrire toutes les plantes—à l'exception des orchidées—ayant un organe souterrain arrondi et volumineux. Cet organe n'est pas toujours un "bulbe" au sens strictement scientifique du terme.

2. Méthodologie
La liste a été compilée en quatre étapes:
- pour chaque genre, les données ont été tirées des révisions taxonomiques les plus récentes et entrées dans une base de données;
- la base de données a été complétée en y incluant tous les noms publiés connus pour chaque genre (en utilisant l'*Index Kewensis*), les publications originales et dans certains cas, des spécimens d'herbiers;
- des notes provisoires ont été établies pour chaque genre, puis revues comme suit:

Cyclamen	Chris Grey-Wilson & Aaron Davis
Galanthus	Aaron Davis
Sternbergia	Brian Mathew & Aaron Davis

Préambule

- la liste finale a été extraite de la base de données et préparée pour l'impression en utilisant *Microsoft Word for Windows version 7*©.

3. Comment utiliser cette liste?

Le but principal de cette liste est de fournir une référence rapide pour vérifier les noms acceptés, les synonymes et la répartition géographique des taxons.

Les "autonymes" y ont été inclus. Les autonymes sont des noms créés automatiquement lorsqu'un nom infraspécifique (inférieur au nom de l'espèce) est ajouté au nom de l'espèce. Par exemple: lorsque *G. bortkewitschianus* est reconnu comme variété de *Galanthus alpinus* (c.-à-d., *Galanthus alpinus* var. *bortkewitschianus*), *Galanthus alpinus* a son épithète répété au même rang que l'addition infraspécifique (c.-à-d. *Galanthus alpinus* var. *alpinus*, qui est l'autonyme). Les autonymes sont inclus dans les première, deuxième et troisième parties. A l'Annexe I, les autonymes sont placés entre parenthèses pour éviter toute confusion.

Les références sont regroupées en trois parties principales:

PREMIÈRE PARTIE: TOUS LES NOMS D'USAGE COURANT

Une liste alphabétique de tous les noms acceptés et synonymes pour les trois genres inclus dans la liste. Dans le première partie, les noms d'auteurs figurent après chaque taxon lorsque le nom apparaît plus d'une fois; par exemple: *Cyclamen cyprium* Glasau et *Cyclamen cyprium* Sibth.

DEUXIÈME PARTIE: NOMS ACCEPTÉS D'USAGE COURANT

Chaque genre est traité dans une liste séparée. Dans chaque liste, les noms acceptés sont présentés par ordre alphabétique et des détails sont donnés sur les synonymes les plus usités et la répartition géographique. Des informations supplémentaires sont fournies sur l'emplacement des aires de répartition dans chaque pays: par exemple, N,S,E,W (ouest), en énumérant les principales îles et en apportant des indications complémentaires sur les taxons introduits. Ces informations figurent également dans la troisième partie, avec des informations supplémentaires sur les taxons introduits.

TROISIÈME PARTIE: LISTE DES PAYS

Les noms acceptés des taxons de chaque genre inclus dans la liste sont classés par ordre alphabétique dans chaque pays de l'aire de répartition.

4. Conventions employées dans les première, deuxième et troisième parties

a) Les noms acceptés sont imprimés **en gras**
les synonymes *en italique*.

b) Lorsque un synonyme apparaît deux fois pour deux noms acceptés différents—par exemple, *Galanthus ikariae* subsp. *latifolius*, à la fois pour **Galanthus platyphyllus** et pour **Galanthus woronowii**—le nom accompagné d'un astérisque est celui l'espèce le plus susceptible d'être rencontrée dans le commerce sous ce nom. Par exemple:

Tous les noms	Noms acceptés
Galanthus ikariae subsp. *latifolius*	**Galanthus platyphyllus**
Galanthus ikariae subsp. *latifolius*	**Galanthus woronowii***

Galanthus ikariae subsp. *snogerupii*   **Galanthus ikariae**

*Espèce le plus susceptible d'être rencontrée dans le commerce (ici, **Galanthus woronowii**)

c) Lorsqu'un nom accepté et un synonyme ont la même épithète mais renvoient à des espèces différentes—par exemple **Cyclamen persicum** et *Cyclamen persicum* (**Cyclamen graecum** subsp. **graecum** forma **graecum**)—le nom accompagné d'un astérisque est celui l'espèce le plus susceptible d'être rencontrée dans le commerce sous ce nom. Par exemple:

Tous les noms   **Noms acceptés**

Cyclamen pentelici   **Cyclamen graecum** subsp.
                                            graecum forma **graecum**
Cyclamen persicum*
Cyclamen persicum Sibth. & Sm   **Cyclamen graecum** subsp.
                                            graecum forma **graecum**

* Espèce le plus susceptible d'être rencontrée dans le commerce (ici, **Cyclamen persicum**)

NB: Dans les exemples b) et c), il convient de vérifier la répartition géographique dans la deuxième partie

d) Une sélection d'hybrides a été incluse dans la liste. Ils se reconnaissent par l'adjonction d'un signe de multiplication ×. Ils sont placés par ordre alphabétique dans les première, deuxième et trosième parties.

5. Décompte des noms retenus pour chaque genre
Cyclamen (acceptés 65, synonymes: 179); *Galanthus* (acceptés 27, synonymes 225); *Sternbergia* (acceptés 9, synonymes 35).

6. Abréviations de titres d'ouvrages, de périodiques et de noms d'auteurs
Les noms d'auteurs sont abrégés selon Brummitt et Powell (1992); les journaux et périodiques selon Lawrence *et al.* (1968) et Bridson et Smith (1991); les titres d'ouvrages selon Stafleu et Cowan (1976 à 1988), avec quelques modifications mineures.

Brummitt, R.K. et Powell, C.E. (1992). *Authors of Plant Names*. Royal Botanic Gardens, Kew.

Lawrence, G.H.M., Günther Buchheim, A.F., Daniels, G.S. et Dolezal, H. (1968). *B-P-H. Botanico-Periodicum-Huntianum*. Pittsburgh: Hunt Botanical Library.

Bridson, G.D.R. et Smith, E.R. (1991). *B-P-H/S Botanico-Periodicum-Huntianum/ Supplementum*. Pittsburgh: Hunt Institute for Botanical Documentation.

Stafleu, F.A. et Cowan, R.S. (1976 - 1988). *Taxonomic Literature*, edn. 2, volumes 1–7. Bohn, Scheltema et Holkema: Utrecht/Antwerpen; dr. W. Junk b.v. Publishers: The Hague/Boston.

Préambule

7. Abréviations, termes botaniques, et mots en latin

Note: les mots *en italique* sont d'origine latine

ambiguous name (nom ambigu) nom donné à différents taxons par différents auteurs, ce qui crée une ambiguïté

anon. anonyme; sans auteur

auct. *auctorum*: d'auteurs

cultivation (culture) obtention de plantes par horticulture ou jardinage, par opposition au prélèvement dans la nature

cultivar spécimen ou groupe de plantes conservant les mêmes caractéristiques distinctives, produites ou conservées (propagées) en culture

descr. *descriptio* description d'une espèce ou d'une autre entité taxonomique

distribution (aire de répartition géographique) région(s) où se trouve les plantes

ed. éditeur

edn. édition (d'un livre ou d'un périodique)

eds. éditeurs

epithet (épithète) dernier mot d'une espèce, d'une sous-espèce ou d'une variété (etc.). Exemple: *lutea* est l'épithète de l'espèce *Sternbergia lutea* et *byzantinus* l'épithète infraspécifique de *Galanthus plicatus* subsp. *byzantinus*

escape (échappée) qualifie une plante qui a quitté l'enceinte de culture (jardin, par exemple) et qu'on retrouve dans la végétation naturelle

ex *ex* d'après; peut être utilisé entre deux noms d'auteurs, dont le second a validement publié le nom d'après les indications ou suggestions du premier

excl. *exclusus* exclu

hort. *hortorum* de jardins (horticole); plante cultivée ou se trouvant dans des jardins horticoles, par opposition à une plante d'origine sauvage

in prep. en préparation

in sched. *in scheda* sur un spécimen d'herbier ou une étiquette

in syn. *in synonymia* en synonymie

incl. incluant

ined. *ineditus* non publié

introduction résultat d'une activité humaine (volontaire ou non) aboutissant à ce qu'une plante non indigène se retrouve dans un pays ou une région

key (clé) système écrit utilisé pour la détermination d'organismes (plantes, par exemple)

leg. *legit* il ramassa; le collecteur

misspelling (faute d'orthographe) nom mal orthographié, par opposition à un nom nouveau ou différent

morphology (morphologie) forme et structure d'un organisme (d'une plante, par exemple)

name causing confusion (nom causant une confusion) nom qui n'est pas utilisé parce qu'il ne peut être assigné sans ambiguïté à un taxon particulier (à une espèce de plante, par exemple)

native (indigène) qualifie un organisme (une plante, par exemple) prospérant naturellement dans un pays ou une région etc.

naturalized (naturalisée) qualifie une plante introduite (voir introduction) ou échappée (voir échappée) qui ressemble à une plante sauvage et qui se propage dans son nouvel environnement

nom. *nomen* nom

nom. ambig. *nomen ambiguum* nom ambigu

nom. cons. prop. *nomen conservandum propositum* nom dont le maintien a été proposé d'après les règles du *International Code of Botanical Nomenclature* (Code international de la nomenclature botanique)

nomenclature branche de la science qui nomme les organismes (les plantes, par exemple)

non *non* pas

only known from cultivation (connue seulement en culture) qualifie une plante qu'on ne trouve pas à l'état sauvage

orthographic variant (variante orthographique) même nom orthographié différemment

pro parte *pro parte* partiellement, en partie

provisional name (nom provisoire) nom donné par anticipation d'une description

sens. *sensu* au sens de; manière dont un auteur interprète ou utilise un nom

sens. lat. *sensu lato* au sens large; un taxon (habituellement une espèce) et tous ses taxons inférieurs (sous-espèce, etc.) et/ou d'autres taxons parfois considérés comme distincts

sic *sic*, utilisé après un mot qui semble faux ou absurde; indique que ce mot est cité textuellement

synonym (synonyme) nom donné à un taxon mais qui ne peut être utilisé parce que ce n'est pas le nom accepté; le ou les synonymes forment la synonymie

taxa pluriel de taxon

taxon unité taxonomique à laquelle on a attribué un nom - genre, espèce, sous-espèce, etc.

8. Noms géographiques

La terminologie des noms de pays est celle des Nations Unies, présentée dans *Country Names, Terminology Bulletin*, août 1995. Nations Unies 347:1-41.

9. Annexes I–VI

Il y a six annexes: (I) liste de tous les noms, comprenant les auteurs et les lieux de publication; (II) clés de détermination des espèces; (III) bibliographie; (IV) liste de synonymes supplémentaires pour *Galanthus nivalis;* (V) résumé des noms acceptés; (VI) une nouvelle combinaison dans *Cyclamen*.

L'Annexe I comprend la liste de tous les noms et des lieux de première publication, indiquant quand et où ils ont été légitimement nommés et par qui. Cette annexe est prévue à l'intention de ceux qui recherchent des informations plus complètes sur ces plantes ornementales.

Les clés de l'Annexe II permettent de déterminer les *Cyclamen*, *Galanthus* (y compris les sous-espèces et variétés) et les *Sternbergia*.

L'Annexe III compile les principales références bibliographiques utilisées. Cette bibliographie permet d'accéder à d'autres informations qui ne sont pas présentées dans les listes, à savoir l'écologie, des détails sur la culture et des descriptions morphologiques.

L'Annexe IV donne la liste des synonymes les plus obscurs de *Galanthus nivalis*, en particulier pour les noms (y compris de nombreux cultivars) issus de la littérature horticole.

Préambule

L'Annexe V indique tous les noms acceptés, ce qui permet à l'utilisateur d'avoir rapidement une vue d'ensemble de chaque genre.

L'Annexe VI donne une nouvelle combinaison dans *Cyclamen*.

LISTA CITES - BULBOS

PREÁMBULO

1. Información general

En 1992, la Conferencia de las Partes en la Convención sobre el Comercio Internacional de Especies Amenazadas de Fauna y Flora Silvestres—CITES (Kyoto, Japón, 1992) se adoptó la *CITES Cactaceae Checklist* como obra de referencia al hacer alusión a las especies de la familia en cuestión. Se trataba de la primera lista de plantas CITES, que fue seguida por la *Orchid Checklist Volume 1* en 1995.

Se ha puesto de manifiesto que estas referencias son un valioso instrumento en las tareas diarias de aplicación de la CITES para las especies de plantas. El apoyo de la Conferencia de las Partes sumado al de los Estados Partes individuales, las instituciones científicas y organizaciones (por ejemplo, la IOS) han hecho posible la preparación y publicación de la *CITES Cactaceae Checklist (segunda edición)*, el *CITES Orchid Checklist Volume 2*, *The CITES Checklist of Succulent Euphorbia Taxa* y la *Bulb Checklist*. Todas estas listas han sido adoptadas por la décima reunión de la Conferencia de las Partes como obras de referencia al hacer alusión a los nombres aceptados de los taxa de que se trata.

La presente lista es el resultado de la cooperación entre el Royal Botanic Gardens, Kew (Reino Unido) y el Ministerio de Agricultura, Gestión de la Naturaleza y Pesca de los Países Bajos. Aprovechando el reciente trabajo realizado sobre los grupos de "bulbos" importantes desde el punto de vista hortícola y altamente comercializados, el Royal Botanic Gardens, Kew contrató a Aaron Davis para que compilase el material de esta publicación y analizase ciertas esferas de problemas. El Ministerio de Agricultura de los Países Bajos proporcionó apoyo financiero para la publicación de la lista.

La *CITES Bulb Checklist* se basa en las recientes revisiones realizadas para *Cyclamen* (Grey-Wilson, 1988 y 1997), *Galanthus* (Davis, 1999) y una nueva sinopsis para el género *Sternbergia* (preparada por B.Mathew y A.P.Davis, publicada por primera vez en la presente lista).

En esta publicación se utiliza la palabra "bulbo" en su acepción corriente, es decir, para representar plantas con yemas subterráneas redondeadas, engrosadas, a excepción de las orquídeas. Evidentemente, no se aplica aquí el significado científico estricto del término "bulbo".

2. Metodología

La lista se compiló en cuatro fases:

* para cada género, se extrajeron datos de las revisiones taxonómicas más recientes y se introdujeron en una base de datos
* la base de datos se amplió incluyendo todos los nombres publicados conocidos para cada género (utilizando *Index Kewensis*), publicaciones originales y, en algunos casos, especímenes de herbario
* se prepararon y revisaron descripciones detalladas para cada género

Cyclamen	Chris Grey-Wilson y Aaron Davis
Galanthus	Aaron Davis

Preámbulo

Sternbergia   Brian Mathew y Aaron Davis

- la lista completa se extrajo de la base de datos y se presentó como material preparado para la cámara utilizando *Microsoft Word for Windows version 7©*.

3. ¿Cómo utilizar esta lista?
La finalidad principal de esta lista es proporcionar una referencia rápida para comprobar los nombres aceptados, la sinonimia y la distribución.

En la presente lista se han introducido autónimos. Los autónimos son nombres que se crean automáticamente cuando se añade un nombre subespecífico (por debajo del nivel de la especie) al nombre de una especie. Por ejemplo: al reconocer *G. bortkewitschianus* como una variedad de *Galanthus alpinus* (dando lugar a *Galanthus alpinus* var. *bortkewitschianus*), *Galanthus alpinus* tiene su epíteto repetido en el mismo rango que la adición subespecífica (resultando en *Galanthus alpinus* var. *alpinus* – que es el autónimo). Los autónimos se incluyen en las Partes I, II y III; en el Anexo I los autónimos se incluyen entre corchetes para evitar confusión.

La referencia se divide en tres partes principales:

PARTE I: TODOS LOS NOMBRES UTILIZADOS NORMALMENTE
Una lista por orden alfabético de todos los nombres y sinónimos aceptados para los tres géneros. En la Parte I el nombre del autor aparece después de cada taxón cuando el nombre aparece dos veces o más, por ejemplo, *Cyclamen cyprium* Glasau y *Cyclamen cyprium* Sibth.

PARTE II: NOMBRES ACEPTADOS UTILIZADOS NORMALMENTE
Listas separadas para cada género. En cada lista se presentan por orden alfabético los nombres aceptados, con información sobre los sinónimos actuales y la distribución. Se incluye información complementaria sobre el área de distribución en cada país, por ejemplo, N, S, E, u O, indicando las islas principales y ofreciendo información sobre las introducciones. Esta información se incluye también en la Parte III, con un suplemento de datos sobre introducciones

PARTE III: LISTA POR PAISES
Los nombres aceptados para todos los géneros incluidos en esta lista se presentan por orden alfabético según el país de distribución.

4. Sistema de presentación utilizado en las Partes I, II y III
a) Los nombres aceptados se presentan en letra **negrita y redonda**.
   Los sinónimos se presentan en letra *bastardilla*.

b) Cuando un sinónimo aparece dos veces, pero se refiere a diferentes nombres aceptados, a saber, *Galanthus ikariae* subsp. *latifolius*, para **Galanthus platyphyllus** y **Galanthus woronowii**, el nombre acompañado de un asteriscose refiere a la especie que se encontrará con mayor probabilidad en el comercio. Por ejemplo:

Todos los nombres	Nombre aceptado
Galanthus ikariae subsp. *latifolius*	**Galanthus platyphyllus**
Galanthus ikariae subsp. *latifolius*	**Galanthus woronowii***

14

Galanthus ikariae subsp. *snogerupii*   **Galanthus ikariae**

*Especie que se encontrará con mayor probabilidad en el comercio (en este ejemplo, **Galanthus woronowii**).

c) Cuando un nombre aceptado y un sinónimo tienen el mismo epíteto, pero se refiere a especies diferentes, a saber, **Cyclamen persicum** y *Cyclamen persicum* (**Cyclamen graecum** subsp. **graecum** forma **graecum**), el nombre acompañado por un asterisco se refiere a la especie que se encontrará con mayor probabilidad en el comercio. Por ejemplo:

Todos los nombres	**Nombre aceptado**
Cyclamen pentelici	**Cyclamen graecum** subsp. **graecum** forma **graecum**
Cyclamen persicum*	
Cyclamen persicum Sibth. & Sm.	**Cyclamen graecum** subsp. **graecum** forma **graecum**

*Especie que se encontrará con mayor probabilidad en el comercio (en este ejemplo, **Cyclamen persicum**).

NB: En los ejemplos b) y c) es preciso efectuar doble verificación en lo que concierne a la distribución, como se indica en la Parte II.

d) Se ha incluido una selección de híbridos y se indican con el signo "×". Se presentan por orden alfabético en la Parte I y al final de las Partes II y III.

5. Número de nombres incluidos para cada género:
Cyclamen (aceptados: 65, sinónimos: 179)*; Galanthus* (aceptados: 27, sinónimos: 225); *Sternbergia* (aceptados: 9, sinónimos: 35).

6. Abreviaciones de títulos de libros, revistas y nombres de autores
Los nombres de los autores se abrevian con arreglo a Brummitt y Powell (1992); los títulos de revistas y periódicos según Lawrence y otros (1968) y Bridson y Smith (1991); y los títulos de los libros según Stafleu y Cowan (1976 a 1988), con algunos pequeños cambios.

Brummitt, R.K. y Powell, C.E. (1992). *Authors of Plant Names.* Royal Botanic Gardens, Kew.

Lawrence, G.H.M., Günther Buchheim, A.F., Daniels, G.S. y Dolezal, H. (1968). *B-P-H. Botanico-Periodicum-Huntianum.* Pittsburgh: Hunt Botanical Library.

Bridson, G.D.R. y Smith, E.R. (1991). *B-P-H/S Botanico-Periodicum-Huntianum/ Supplementum.* Pittsburgh: Hunt Institute for Botanical Documentation.

Stafleu, F.A. y Cowan, R.S. (1976 - 1988). *Taxonomic Literature*, edn. 2, volumes 1–7. Bohn, Scheltema and Holkema: Utrecht/Antwerpen; dr. W. Junk b.v. Publishers: The Hague/Boston.

7. Abreviaciones, términos botánicos, y expresiones latinas

Nota: las expresiones latinas aparecen en *bastardilla*

ambiguous name (nombre ambiguo) un nombre utilizado por distintos autores para diferentes taxa, de manera que da motivo a confusión

anon. Anonymous; autor desconocido

auct. *auctorum* de autores

cultivation (cultivo) el cultivo de plantas mediante horticultura o jardinería; no se ha recolectado inmediantemente del medio silvestre

cultivar un ejemplar, o una agrupación de plantas, que tienen los mismos rasgos característicos, que ha sido producido o se mantiene (reproduce) en cultivo

descr. *descriptio* la descripción de una especie o de otra unidad taxonómica

distribution (distribución) donde se encuentran las plantas (geográfica)

ed. editor

edn. edición (libro o revista)

eds. editores

eptithet (epíteto) la última palabra de una especie, subespecie o variedad (etc.), por ejemplo: *lutea* es el epíteto de la especie *Sternbergia lutea* y *byzantinus* es el epíteto subespecífico de *Galanthus plicatus* subsp. *byzantinus*

escape (volverse silvestre) una planta que ha sobrepasado los límites del cultivo (p.e.: un jardín) y prospera en la naturaleza

ex *ex* después, puede utilizarse entre los nombres de dos autores, el segundo de los cuales publicó el nombre indicado o sugerido por el primero

excl. *exclusus* excluida

hort. *hortorum* de jardines (horticultura); cultivadas o prosperan en jardines; no se trata de una planta silvestre

incl. inclusive

in prep. en preparación

in sched. *in scheda* en un espécimen de herbario o etiqueta

in syn. *in synonymia* en sinonimia

ined. *ineditus* : inédito

introduction (introducción) una planta que ocurre en un país, o en cualquier otra localidad, debido a la influencia antropogénica (intencionalmente o al azar); cualquier planta que no es nativa

key (clave) un sistema escrito utilizado para la identificación de organismos (p.e.: plantas)

leg. *legit* el recolector; el coleccionista

misspelling (error de ortografía) un nombre que se ha escrito incorrectamente; no se trata de un nombre nuevo o diferente

morphology (morfología) la forma y estructura de un organismo (p.e.: una planta)

nombre que crea confusión: un nombre que no se usa, ya que su utilización crearía confusión

name causing confusion (nombre de dudosa semejanza) un nombre que no se usa, ya que no puede asignarse a un determinado taxón sin crear confusión (p.e.: una especie de planta)

native (nativo) un organismo (p.e.: una planta) que prospera naturalmente en un país o región, etc.

naturalized (naturalizada) una planta que ha sido introducida (véase introducción) o se ha vuelto silvestre (véase volverse silvestre) pero que parece una planta silvestre y se reproduce por sí misma en su nuevo medio

nom. *nomen* nombre

nom. ambig. *nomen ambiguum* nombre ambiguo

nom. cons. prop. *nomen conservandum propositum* nombre propuesto para la conservación con arreglo a lo dispuesto en el Código Internacional de Nomenclatura Botánica (ICBN)

nomenclature (nomenclatura) parte de la ciencia que se ocupa de atribuir nombres a organismos (p.e.: plantas)

non *non* no

only known from cultivation (solo se conoce en cultivo) una planta que no ocurre en el medio silvestre, únicamente en cultivo

orthographic variant (variante ortográfica) una alternativa ortográfica del mismo nombre

pro parte *pro parte* : parcialmente, en parte

provisional name (nombre provisional) nombre asignado temporalmente hasta que se disponga de una descripción válida

sens. *sensu* en el sentido de; la forma en que un autor interpreta o utiliza un nombre

sens. lat. *sensu lato* en sentido generalizado, un taxón (normalmente una especie) y todos sus taxa subordinados (p.e.: subspecies) y/o otros taxa a veces considerados como distintos

sic *sic* utilizado después de una palabra que pudiera parecer inexacta o absurda, para dar a entender que es textual

synonym (sinónimo) un nombre que se aplica a un taxón pero que no puede utilizarse ya que no es un nombre aceptado – el sinónimo o los sinónimos forman la sinonimia

taxa plural de taxón

taxon (taxón) una determinada unidad de clasificación, p.e.: género, especie, subespecie

8. Areas geográficas

Para los nombres de los países se ha seguido la referencia oficial de las Naciones Unidas. *Terminology Bulletin* August 1995. United Nations 347:1–41.

9. Anexos I-VI

Hay seises anexos: (I) lista de todos los nombres, indicando el autor y el lugar de publicación; (II) claves para las especies; (III) bibliografía; (IV) lista de otros sinónimos de *Galanthus nivalis*; (V) resumen de los nombres aceptados; (VI) una nueva combinación en *Cyclamen*.

El Anexo I es una lista de todos los nombres y su lugar original de publicación, p.e.: cuando y donde se acuño su nombre legítico y por quién. Está destinado a aquellas personas que desean disponer de mayor información sobre estas plantas ornamentales.

El Anexo II contiene claves que constituyen un instrumento para la identificación de *Cyclamen*, *Galanthus* (inclusive subspecies y variedades) y *Sternbergia*.

El Anexo III es una bibliografía de las principales referencias utilizadas en esta lista; permiten acceder a otra información que no figura en esta publicación, como la ecología, datos sobre el cultivo y descripciones sobre la morfología.

En el Anexo IV figura una lista de los sinónimos más confusos de *Galanthus nivalis*, concretamente para aquellos nombres (inclusive muchos cultivares) que aparecen en la literatura hortícola.

Preámbulo

En el Anexo V se ofrece una numeración de todos los nombres aceptados, permitiendo al usuario examinar de un vistazo el contenido dentro de cada género.

En el Anexo VI da una nueva combinación en *Cyclamen*.

PART I: ALL NAMES IN CURRENT USE
Ordered alphabetically on all names for the genera:

Cyclamen, *Galanthus* and *Sternbergia*

PREMIERE PARTIE: LISTE ALPHABÉTIQUE DE TOUS LES NOMS
D'USAGE COURANT
Par ordre alphabétique de tous les noms

Cyclamen, *Galanthus* et *Sternbergia*

PARTE I: TODOS LOS NOMBRES UTILIZADOS NORMALMENTE
Presentados orden alfabético de todos los nombres para los géneros:

Cyclamen, *Galanthus* y *Sternbergia*

ALPHABETICAL LISTING OF ALL NAMES FOR THE GENERA:
Cyclamen, *Galanthus* and *Sternbergia*

LISTES ALPHABÉTIQUES DE TOUS LES NOMS POUR LES GENERE:
Cyclamen, *Galanthus* et *Sternbergia*

PRESENTACION POR ORDEN ALFABÉTICO DE TODOS LOS NOMBRES PARA LOS GÉNEROS:
Cyclamen, *Galanthus* y *Sternbergia*

ALL NAMES TOUS LES NOMS TODOS LOS NOMBRES	ACCEPTED NAME NOM RECONNU NOMBRES ACEPTADOS
Amaryllis aetnensis	**Sternbergia colchiciflora**
Amaryllis citrina	**Sternbergia colchiciflora**
Amaryllis clusiana	**Sternbergia clusiana**
Amaryllis colchiciflora	**Sternbergia colchiciflora**
Amaryllis lutea L.	**Sternbergia lutea**
Amaryllis lutea M.Bieb.	**Sternbergia fischeriana**
Amaryllis vernalis	**Sternbergia fischeriana**
Chianthemum elwesii	**Galanthus elwesii**
Chianthemum graecum	**Galanthus elwesii**
Chianthemum nivale	**Galanthus nivalis**
Chianthemum olgae	**Galanthus reginae-olgae**
Chianthemum plicatum	**Galanthus plicatus** subsp. **plicatus**
Cyclamen abchasicum	**Cyclamen coum** subsp. **caucasicum**
Cyclamen adzharicum	**Cyclamen coum** subsp. **caucasicum**
Cyclamen aedirhizum	**Cyclamen hederifolium** var. **hederifolium** forma **hederifolium**
Cyclamen aegineticum	**Cyclamen graecum** subsp. **graecum** forma **graecum**
Cyclamen aestivum	**Cyclamen purpurascens** forma **purpurascens**
Cyclamen africanum	
Cyclamen albidum	**Cyclamen persicum** var. **persicum** forma **albidum**
Cyclamen albiflorum	**Cyclamen hederifolium** var. **hederifolium** forma **albiflorum**
Cyclamen aleppicum	**Cyclamen persicum** var. **persicum** forma **persicum**
Cyclamen aleppicum subsp. *puniceum*	**Cyclamen persicum** var. **persicum** forma **puniceum**
Cyclamen algeriense	**Cyclamen africanum**
Cyclamen alpinum hort. Dammann ex Sprenger	**Name causing confusion**
Cyclamen alpinum sens. Turrill, pro parte	**Cyclamen trochopteranthum** forma **trochopteranthum**
Cyclamen alpinum var. *album*	**Name causing confusion**
Cyclamen ambiguum	**Cyclamen africanum**
Cyclamen angulare	**Cyclamen hederifolium** var. **hederifolium** forma **hederifolium**

*For explanation see page 2, point 4
*Voir les explications page 8, point 4
*Para mayor explicación, véase la página 14, point 4

ALL NAMES	ACCEPTED NAME
Cyclamen antilochium	**Cyclamen persicum** var. **persicum** forma **persicum**
Cyclamen apiculatum	**Cyclamen coum** subsp. **coum** forma **coum**
Cyclamen atkinsii Glasau	**Cyclamen coum** subsp. **coum** forma **coum**
Cyclamen atkinsii hort.	**Cyclamen coum** subsp. **caucasicum***
Cyclamen × **atkinsii**	
Cyclamen autumnale	**Name of uncertain affinity**
Cyclamen baborense	**Cyclamen repandum** subsp. **repandum** var. **baborense**
Cyclamen balearicum	
Cyclamen breviflorum	**Cyclamen purpurascens** forma **purpurascens**
Cyclamen brevifrons	**Cyclamen coum** subsp. **coum** forma **coum**
Cyclamen calcareum	**Cyclamen coum** subsp. **caucasicum**
Cyclamen caucasicum	**Cyclamen coum** subsp. **caucasicum**
Cyclamen cilicium*	
Cyclamen cilicicum	**Cyclamen cilicium**
Cyclamen cilicium Hildebr.	**Name of uncertain affinity**
Cyclamen cilicium forma **album**	
Cyclamen cilicium forma **cilicium**	
Cyclamen cilicium var. *alpinum*	**Cyclamen intaminatum**
Cyclamen cilicium var. *intaminatum*	**Cyclamen intaminatum**
Cyclamen cilicium var. [*sic*]	**Cyclamen intaminatum**
Cyclamen circassicum	**Cyclamen coum** subsp. **caucasicum**
Cyclamen clusii	**Cyclamen purpurascens** forma **purpurascens**
Cyclamen colchicum	
Cyclamen commutatum	**Cyclamen africanum**
Cyclamen cordifolium	**Name of uncertain affinity**
Cyclamen coum	
Cyclamen coum sens. Rchb.	**Name of uncertain affinity**
Cyclamen coum subsp. *alpinum*	**Name causing confusion**
Cyclamen coum subsp. **caucasicum**	
Cyclamen coum subsp. **coum** forma **albissimum**	
Cyclamen coum subsp. **coum** forma **coum**	
Cyclamen coum subsp. **coum** forma **pallidum**	
Cyclamen coum subsp. **elegans**	
Cyclamen coum subsp. *hiemale*	**Cyclamen coum** subsp. **coum** forma **coum**
Cyclamen coum var. *abchasicum*	**Cyclamen coum** subsp. **caucasicum**
Cyclamen coum var. *caucasicum*	**Cyclamen coum** subsp. **caucasicum**
Cyclamen coum var. *ibericum*	**Cyclamen coum** subsp. **caucasicum**
Cyclamen crassifolium	**Name of uncertain affinity**
Cyclamen creticum	
Cyclamen creticum forma **creticum**	
Cyclamen creticum forma **pallide-roseum**	
Cyclamen cyclaminus	**Cyclamen hederifolium** var. **hederifolium** forma **hederifolium**
Cyclamen cyclophyllum	**Cyclamen purpurascens** forma **purpurascens**
Cyclamen cyprium *	
Cyclamen cyprium Glasau	**? Cyclamen graecum** subsp. **anatolicum**

*For explanation see page 2, point 4
*Voir les explications page 8, point 4
*Para mayor explicación, véase la página 14, point 4   21

ALL NAMES	ACCEPTED NAME
Cyclamen cyprium Sibth.	**Cyclamen persicum** var. **persicum** forma **persicum***
Cyclamen cypro-graecum	**Cyclamen graecum** subsp. **anatolicum**
Cyclamen deltoideum	**Cyclamen purpurascens** forma **purpurascens**
Cyclamen × drydeniae	
Cyclamen durostoricum	**Cyclamen coum** subsp. **coum** forma **coum**
Cyclamen elegans	**Cyclamen coum** subsp. **elegans**
Cyclamen eucardium	**Cyclamen repandum** subsp. **peloponnesiacum** var. **vividum**
Cyclamen europaeum L., pro parte	**Cyclamen purpurascens** forma **purpurascens**
Cyclamen europaeum L., pro parte	**Cyclamen hederifolium** var. **hederifolium** forma **hederifolium***
Cyclamen europaeum L., pro parte	**Cyclamen repandum** subsp. **repandum** var. **repandum** forma **repandum***
Cyclamen europaeum Pall.	**Cyclamen coum** subsp. **coum** forma **coum**
Cyclamen europaeum Savi	**Cyclamen purpurascens** forma **purpurascens**
Cyclamen europaeum sens. Aiton	**Cyclamen purpurascens** forma **purpurascens**
Cyclamen europaeum sens. Albov	**Cyclamen colchicum**
Cyclamen europaeum sens. Mill.	**Name causing confusion**
Cyclamen europaeum subsp. *orbiculatum*	**Cyclamen purpurascens** forma **purpurascens**
Cyclamen europaeum subsp. *orbiculatum* var. *immaculatum*	**Cyclamen purpurascens** forma **purpurascens**
Cyclamen europaeum subsp. *ponticum*	**Cyclamen colchicum**
Cyclamen europaeum Sm.	**Name of uncertain affinity**
Cyclamen europaeum var. *caucasicum*	**Cyclamen coum** subsp. **caucasicum**
Cyclamen europaeum var. *colchicum*	**Cyclamen colchicum**
Cyclamen europaeum var. *ponticum*	**Cyclamen colchicum**
Cyclamen europaeum var. *typicum*	**Cyclamen purpurascens** forma **purpurascens**
Cyclamen fatrense	**Cyclamen purpurascens** forma **purpurascens**
Cyclamen ficariifolium	**Cyclamen repandum** subsp.**repandum** var. **repandum** forma **repandum**
Cyclamen floridum	**Cyclamen purpurascens** forma **purpurascens**
Cyclamen gaidurowryssii	**Cyclamen graecum** subsp. **graecum** forma **graecum**
Cyclamen gaydurowryssii (orthographical variant)	**Cyclamen graecum** subsp. **graecum** forma **graecum**
Cyclamen graecum	
Cyclamen graecum subsp. **anatolicum**	
Cyclamen graecum subsp. *candicum*	**Cyclamen graecum** subsp. **mindleri**
Cyclamen graecum subsp. **graecum** forma **album**	
Cyclamen graecum subsp. **graecum** forma **graecum**	
Cyclamen graecum subsp. **mindleri**	
Cyclamen hastatum	**Cyclamen purpurascens** forma **purpurascens**

*For explanation see page 2, point 4
*Voir les explications page 8, point 4
*Para mayor explicación, véase la página 14, point 4

ALL NAMES

Cyclamen hederaceum Sieber ex Steud.

Cyclamen hederifolium*
Cyclamen hederifolium Kotschy

Cyclamen hederifolium Sibth. & Sm.

Cyclamen hederifolium Sims

Cyclamen hederifolium Willd.
Cyclamen hederifolium subsp. *balearicum*
Cyclamen hederifolium subsp. *creticum*
Cyclamen hederifolium subsp. *romanum*

Cyclamen hederifolium var. confusum
Cyclamen hederifolium var. hederifolium
forma albiflorum
Cyclamen hederifolium var. hederifolium
forma hederifolium
Cyclamen hiemale

Cyclamen × hildebrandii
Cyclamen holochlorum

Cyclamen hyemale

Cyclamen ibericum
Cyclamen ilicetorum

Cyclamen immaculatum
Cyclamen indicum
Cyclamen insulare

Cyclamen intaminatum
Cyclamen intermedium
Cyclamen jovis
Cyclamen kusnetzovii

Cyclamen latifolium

Cyclamen libanoticum
Cyclamen libanoticum subsp. *pseudibericum*

Cyclamen lilacinum

Cyclamen linaerifolium

Cyclamen littorale

Cyclamen lobospilum

Cyclamen macrophyllum
Cyclamen macropus
Cyclamen maritimum
Cyclamen × marxii

ACCEPTED NAME

Cyclamen persicum var. persicum
forma persicum

Cyclamen persicum var. persicum
forma persicum
Cyclamen hederifolium var.
hederifolium forma hederifolium*
Cyclamen repandum subsp. repandum
var. repandum forma repandum
Name of uncertain affinity
Cyclamen balearicum
Cyclamen creticum forma creticum
Cyclamen hederifolium var.
hederifolium forma hederifolium

Cyclamen coum subsp. coum forma
coum

Cyclamen purpurascens forma
purpurascens
Cyclamen coum subsp. coum forma
coum
Cyclamen coum subsp. caucasicum
Cyclamen repandum subsp.repandum
forma. repandum
Name of uncertain affinity
Name causing confusion
Cyclamen hederifolium var.
hederifolium forma hederifolium

Name of uncertain affinity
Name of uncertain affinity
Cyclamen coum subsp. coum forma
coum
Cyclamen persicum var. persicum
forma persicum

Cyclamen pseudibericum forma
pseudibericum
Cyclamen purpurascens forma
purpurascens
Cyclamen hederifolium var.
hederifolium forma hederifolium
Cyclamen purpurascens forma
purpurascens
Cyclamen repandum subsp. repandum
var. repandum forma repandum
Name of uncertain affinity
Name of uncertain affinity
Cyclamen graecum subsp. anatolicum
Name of uncertain affinity

*For explanation see page 2, point 4
*Voir les explications page 8, point 4
*Para mayor explicación, véase la página 14, point 4   23

Part I: All Names / Tous les Noms / Todos los Nombres

ALL NAMES	ACCEPTED NAME
Cyclamen × meiklei	
Cyclamen miliarakesii	Cyclamen graecum subsp. graecum forma graecum
Cyclamen mirabile	
Cyclamen mirabile forma mirabile	
Cyclamen mirabile forma niveum	
Cyclamen neapolitanum sens. Boiss.	Cyclamen cyprium
Cyclamen neapolitanum sens. Duby	Cyclamen africanum
Cyclamen neapolitanum Ten.	Cyclamen hederifolium var. hederifolium forma hederifolium*
Cyclamen numidicum	Cyclamen africanum
Cyclamen officinale	Name of uncertain affinity
Cyclamen orbiculatum	Cyclamen coum subsp. coum forma coum
Cyclamen orbiculatum var. alpinum	Cyclamen trochopteranthum forma trochopteranthum
Cyclamen orbiculatum var. coum	Cyclamen coum subsp. coum forma coum
Cyclamen pachylobum	Cyclamen africanum
Cyclamen parviflorum	
Cyclamen parviflorum var. parviflorum	
Cyclamen parviflorum var. subalpinum	
Cyclamen pentelici	Cyclamen graecum subsp. graecum forma graecum
Cyclamen persicum*	
Cyclamen persicum sens. Sibth. & Sm.	Cyclamen graecum subsp. graecum forma graecum
Cyclamen persicum subsp. eupersicum	Cyclamen persicum var. persicum forma persicum
Cyclamen persicum subsp. mindleri	Cyclamen graecum subsp. mindleri
Cyclamen persicum var. autumnale	
Cyclamen persicum var. persicum forma albidum	
Cyclamen persicum var. persicum forma persicum	
Cyclamen persicum var. persicum forma puniceum	
Cyclamen persicum var. persicum forma roseum	Provisional name
Cyclamen poli	Cyclamen hederifolium var. hederifolium forma hederifolium
Cyclamen ponticum	Cyclamen colchicum
Cyclamen pseudibericum	
Cyclamen pseudibericum forma pseudibericum	
Cyclamen pseudibericum forma roseum	
Cyclamen pseudograecum	Cyclamen graecum subsp. mindleri
Cyclamen pseudomaritimum	Cyclamen graecum subsp. anatolicum
Cyclamen punicum	Cyclamen persicum var. persicum forma persicum
Cyclamen purpurascens	
Cyclamen purpurascens forma album	
Cyclamen purpurascens forma purpurascens	
Cyclamen purpurascens subsp. immaculatum	Cyclamen purpurascens forma purpurascens
Cyclamen purprurascens subsp. ponticum	Cyclamen colchicum
Cyclamen pyrolifolium	Cyclamen persicum var. persicum forma persicum

*For explanation see page 2, point 4
*Voir les explications page 8, point 4
*Para mayor explicación, véase la página 14, point 4

24

ALL NAMES	ACCEPTED NAME
Cyclamen rarinaevum	Cyclamen repandum subsp. repandum var. **repandum** forma **repandum**
Cyclamen repandum*	
Cyclamen repandum	Cyclamen balearicum
Cyclamen repandum subsp. *atlanticum*	Cyclamen repandum subsp. repandum var. **baborense**
Cyclamen repandum subsp. *balearicum*	Cyclamen balearicum
Cyclamen repandum subsp. *peloponnesiacum* forma *peloponnesiacum*	Cyclamen repandum subsp. peloponnesiacum var. **peloponnesiacum**
Cyclamen repandum subsp. *peloponnesiacum* forma *vividum*	Cyclamen repandum subsp. peloponnesiacum var. **vividum**
Cyclamen repandum subsp. peloponnesiacum var. **peloponnesiacum**	
Cyclamen repandum subsp. peloponnesiacum var. **vividum**	
Cyclamen repandum subsp. repandum var. repandum forma **album**	
Cyclamen repandum subsp. repandum var. repandum forma **repandum**	
Cyclamen repandum subsp. repandum var. **baborense**	
Cyclamen repandum subsp. **rhodense**	
Cyclamen repandum var. *creticum*	Cyclamen creticum forma **creticum**
Cyclamen repandum var. *rhodense*	Cyclamen repandum subsp. **rhodense**
Cyclamen repandum var. *stenopetalum*	Cyclamen balearicum
Cyclamen retroflexum	Cyclamen purpurascens forma **purpurascens**
Cyclamen rhodium	Cyclamen repandum subsp. **rhodense**
Cyclamen rohlfsianum	
Cyclamen romanum	Cyclamen hederifolium var. **hederifolium** forma **hederifolium**
Cyclamen rotundifolium	Name of uncertain affinity
Cyclamen sabaudum	Cyclamen hederifolium var. **hederifolium** forma **hederifolium**
Cyclamen saldense	Cyclamen africanum
Cyclamen × saundersiae	
Cyclamen × schwarzii	
Cyclamen somalense	
Cyclamen spectabile	Cyclamen repandum subsp. peloponnesiacum var. **vividum**
Cyclamen stenopetalum	Cyclamen repandum subsp. peloponnesiacum var. **vividum**
Cyclamen subhastatum	Cyclamen hederifolium var. **hederifolium** forma **hederifolium**
Cyclamen subrotundum	Cyclamen africanum
Cyclamen tauricum	Name of uncertain affinity
Cyclamen trochopteranthum	
Cyclamen trochopteranthum forma **leucanthum**	
Cyclamen trochopteranthum forma **trochopteranthum**	
Cyclamen tunetanum	Cyclamen persicum var. **persicum** forma **persicum**
Cyclamen umbratile	Cyclamen purpurascens forma **purpurascens**

*For explanation see page 2, point 4
*Voir les explications page 8, point 4
*Para mayor explicación, véase la página 14, point 4

ALL NAMES	ACCEPTED NAME
Cyclamen utopicum	Cyclamen persicum var. persicum forma persicum
Cyclamen variegatum	Cyclamen purpurascens forma purpurascens
Cyclamen velutinum	Cyclamen graecum subsp. graecum forma graecum
Cyclamen venustum	Cyclamen africanum
Cyclamen vernale hort.	Cyclamen coum subsp. coum forma coum*
Cyclamen vernale Mill.	Cyclamen persicum var. persicum forma persicum
Cyclamen vernale sens. O.Schwarz	Cyclamen repandum subsp. repandum var. repandum forma repandum
Cyclamen vernum Lobel ex Cambess.	Cyclamen balearicum
Cyclamen vernum Lobel ex Rchb.	Cyclamen repandum subsp. repandum var. repandum forma repandum*
Cyclamen vernum Sweet	Cyclamen coum subsp. caucasicum
Cyclamen vernum forma *alpinum*	Name causing confusion
Cyclamen vernum var. *caucasicum*	Cyclamen coum subsp. caucasicum
Cyclamen vernum var. *hiemale* forma *alpinum*	Name causing confusion
Cyclamen vernum var. *hiemale* forma *pseudocoum*	Cyclamen coum subsp. coum forma coum
Cyclamen × wellensiekii	
Cyclamen × whiteae	
Cyclamen zonale	Cyclamen coum subsp. caucasicum
Cyclaminos graeca	Cyclamen graecum subsp. graecum forma graecum
Cyclaminos miliarakesii	Cyclamen graecum subsp. graecum forma graecum
Cyclaminos mindleri	Cyclamen graecum subsp. mindleri
Cyclaminum vernum	Name of uncertain affinity
Cyclaminus coa	Cyclamen coum subsp. coum forma coum
Cyclaminus europaea	Ambiguous name
Cyclaminus europaeus [*sic*]	Ambiguous name
Cyclaminus graeca	Cyclamen graecum subsp. graecum forma graecum
Cylaminus neopolitana	Cyclamen hederifolium var. hederifolium forma hederifolium
Cyclaminus persica	Cyclamen persicum var. persicum forma persicum
Cyclaminus repanda	Cyclamen repandum subsp. repandum forma repandum
Galanthus alexandrii	Galanthus nivalis
Galanthus allenii	Galanthus × allenii
Galanthus × allenii	
Galanthus alpinus	
Galanthus alpinus var. alpinus	
Galanthus alpinus var. bortkewitschianus	
Galanthus angustifolius	
Galanthus atkinsii ⊕	Galanthus nivalis
Galanthus bortkewitschianus	Galanthus alpinus var. borkewitschianus
Galanthus bulgaricus	Galanthus elwesii
Galanthus byzantinus	Galanthus plicatus subsp. byzantinus

*For explanation see page 2, point 4
*Voir les explications page 8, point 4
*Para mayor explicación, véase la página 14, point 4

ALL NAMES	ACCEPTED NAME
Galanthus byzantinus subsp. *brauneri*	**Galanthus plicatus** subsp. **byzantinus**
Galanthus byzantinus subsp. *saueri*	**Galanthus plicatus** subsp. **byzantinus**
Galanthus byzantinus subsp. *tughrulii*	**Galanthus plicatus** subsp. **byzantinus**
Galanthus cabardensis	**Galanthus lagodechianus**
Galanthus caspius	**Galanthus transcaucasicus**
Galanthus caucasicus	**Galanthus alpinus** var. **alpinus**
Galanthus caucasicus hort.	**Galanthus elwesii**
Galanthus caucasicus var. *hiemalis*	**Galanthus elwesii**
Galanthus cilicicus*	
Galanthus cilicicus auct. non Baker	**Galanthus peshmenii***
Galanthus cilicicus auct. non Baker	**Galanthus rizehensis**
Galanthus cilicicus subsp. *caucasicus*	**Galanthus alpinus** var. **alpinus**
Galanthus clusii	**Galanthus plicatus** subsp. **plicatus**
Galanthus corcynensis [*sic*]	**Galanthus reginae-olgae** subsp. **reginae-olgae**
Galanthus corcyrensis	**Galanthus reginae-olgae** subsp. **reginae-olgae**
Galanthus corcyrensis (*praecox*)	**Galanthus reginae-olgae** subsp. **reginae-olgae**
Galanthus elsae	**Galanthus reginae-olgae** subsp. **reginae-olgae**
Galanthus elwesii	
Galanthus elwesii subsp. *akmanii*	**Galanthus elwesii**
Galanthus elwesii subsp. *baytopii*	**Galanthus elwesii**
Galanthus elwesii subsp. *melihae*	**Galanthus elwesii**
Galanthus elwesii subsp. *minor*	**Galanthus gracilis**
Galanthus elwesii subsp. *tuebitaki*	**Galanthus elwesii**
Galanthus elwesii subsp. *wagenitzii*	**Galanthus elwesii**
Galanthus elwesii subsp. *yayintaschii*	**Galanthus elwesii**
Galanthus elwesii var. *globosus*	**Galanthus elwesii**
Galanthus elwesii var. *maximus*	**Galanthus elwesii**
Galanthus elwesii var. *monostictus*	**Galanthus elwesii**
Galanthus elwesii var. *platyphyllus*	**Galanthus elwesii**
Galanthus elwesii var. *reflexus*	**Galanthus gracilis**
Galanthus elwesii var. *robustus*	**Galanthus elwesii**
Galanthus elwesii var. *stenophyllus*	**Galanthus gracilis**
Galanthus elwesii var. *whittallii*	**Galanthus elwesii**
Galanthus fosteri	
Galanthus fosteri var. *antepensis*	**Galanthus fosteri**
Galanthus glaucescens	**Galanthus rizehensis**
Galanthus globosus	**Galanthus elwesii**
Galanthus gracilis	
Galanthus gracilis subsp. *baytopii*	**Galanthus elwesii**
Galanthus graecus	**Galanthus elwesii**
Galanthus graecus auct. non Orph. ex Boiss.	**Galanthus gracilis***
Galanthus graceus forma *gracilis*	**Galanthus gracilis**
Galanthus graceus forma *maximus*	**Galanthus elwesii**
Galanthus graecus var. *maximus*	**Galanthus elwesii**
Galanthus × grandiflorus	**Garden hybrid**
Galanthus grandis	**Galanthus alpinus** var. **alpinus**
Galanthus ikariae*	
Galanthus ikariae auct. non Baker, pro parte	**Galanthus woronowii**
Galanthus ikariae subsp. *latifolius* pro parte	**Galanthus platyphyllus**
Galanthus ikariae subsp. *latifolius* pro parte	**Galanthus woronowii***
Galanthus ikariae subsp. *snogerupii*	**Galanthus ikariae**

***For explanation see page 2, point 4**
***Voir les explications page 8, point 4**
***Para mayor explicación, véase la página 14, point 4**   27

ALL NAMES	ACCEPTED NAME
Galanthus imperati ✤	**Galanthus nivalis**
Galanthus imperati forma *australis*	**Galanthus reginae-olgae** subsp. **reginae-olgae**
Galanthus kemulariae	**Galanthus lagodechianus**
Galanthus ketzkhovelii	**Galanthus lagodechianus**
Galanthus koenenianus	
Galanthus krasnovii	
Galanthus krasnovii subsp. *maculatus*	**Galanthus krasnovii**
Galanthus lagodechianus	
Galanthus latifolius auct. non Rupr.	**Galanthus woronowii**
Galanthus latifolius Rupr.	**Galanthus platyphyllus**
Galanthus latifolius Salisb.	**Galanthus plicatus** subsp. **plicatus**
Galanthus latifolius forma *allenii*	**Galanthus × allenii**
Galanthus latifolius forma *fosteri*	**Galanthus fosteri**
Galanthus latifolius forma *typicus*	**Galanthus platyphyllus**
Galanthus latifolius [var.] *rizaensis*	**Galanthus rizehensis**
Galanthus latifolius var. *rizehensis*	**Galanthus rizehensis**
Galanthus maximus	**Galanthus elwesii**
Galanthus × maximus	**Galanthus × grandiflorus**
Galanthus melihae	**Galanthus elwesii**
Galanthus montana	**Galanthus nivalis**
Galanthus nivalis*	
Galanthus nivalis 'Flore Pleno'	
Galanthus nivalis sens. Ledeb.	**Galanthus alpinus** var. **alpinus**
Galanthus nivalis forma *octobrinus*	**Galanthus reginae-olgae** subsp. **reginae-olgae**
Galanthus nivalis forma *pictus*	**Galanthus nivalis**
Galanthus nivalis forma *pleniflorus*	**Galanthus nivalis 'Flore Pleno'**
Galanthus nivalis subsp. *allenii*	**Galanthus × allenii**
Galanthus nivalis subsp. *angustifolius*	**Galanthus angustifolius**
Galanthus nivalis subsp. *byzantinus*	**Galanthus plicatus** subsp. **byzantinus**
Galanthus nivalis subsp. *caucasicus*	**Galanthus alpinus** var. **alpinus**
Galanthus nivalis subsp. *cilicicus*	**Galanthus cilicicus**
Galanthus nivalis subsp. *cilicicus* auct. non (Baker) Gottl.-Tann.	**Galanthus peshmenii**
Galanthus nivalis subsp. *elwesii*	**Galanthus elwesii**
Galanthus nivalis subsp. *graecus*	**Galanthus elwesii**
Galanthus nivalis subsp. *humboldtii*	**Galanthus nivalis**
Galanthus nivalis subsp. *imperati* ✤	**Galanthus nivalis**
Galanthus nivalis subsp. *plicatus*	**Galanthus plicatus** subsp. **plicatus**
Galanthus nivalis subsp. *reginae-olgae*	**Galanthus reginae-olgae** subsp. **reginae-olgae**
Galanthus nivalis subsp. *subplicatus*	**Galanthus nivalis**
Galanthus nivalis [var.] *atkinsii* J.Allen ✤	**Galanthus nivalis**
Galanthus nivalis var. *atkinsii* Mallet ✤	**Galanthus nivalis**
Galanthus nivalis var. *carpaticus*	**Galanthus nivalis**
Galanthus nivalis var. *caspius*	**Galanthus transcaucasicus**
Galanthus nivalis var. *caucasicus* (Baker) Beck	**Galanthus alpinus** var. **alpinus**
Galanthus nivalis var. *caucasicus* (Baker) Fomin	**Galanthus alpinus** var. **alpinus**
Galanthus nivalis var. *caucasicus* (Baker) J. Phillippow	**Galanthus alpinus** var. **alpinus**
Galanthus nivalis var. *corcyrensis* (Beck) Halácasy	**Galanthus reginae-olgae** subsp. **reginae-olgae**
Galanthus nivalis [var.] *corcyrensis* hort. ex Leichtlin	**Galanthus reginae-olgae** subsp. **reginae-olgae**

*For explanation see page 2, point 4
*Voir les explications page 8, point 4
*Para mayor explicación, véase la página 14, point 4

28

ALL NAMES

ACCEPTED NAME

ALL NAMES	ACCEPTED NAME
Galanthus [*nivalis*] var. *elsae*	**Galanthus reginae-olgae** subsp. **reginae-olgae**
Galanthus nivalis var. *europaeus* forma *corcyrensis*	**Galanthus reginae-olgae** subsp. **reginae-olgae**
Galanthus nivalis var. *europaeus* forma *hololeucus*	**Galanthus nivalis**
Galanthus nivalis var. *europaeus* forma *hortensis*	**Galanthus nivalis**
Galanthus nivalis var. *europaeus* forma *olgae*	**Galanthus reginae-olgae** subsp. **reginae-olgae**
Galanthus nivalis var. *europaeus* forma *scharloki* [*sic*] ✥	**Galanthus nivalis**
Galanthus nivalis var. *grandior*	**Galanthus nivalis**
Galanthus nivalis var. *hololeucus*	**Galanthus nivalis**
Galanthus nivalis var. *hortensis*	**Galanthus nivalis**
Galanthus nivalis var. *imperati* ✥	**Galanthus nivalis**
Galanthus nivalis var. *major*	**Galanthus alpinus** var. **alpinus**
Galanthus nivalis var. *major* sens. Fiori	**Galanthus nivalis**
Galanthus nivalis var. *majus* [*sic*]	**Galanthus nivalis**
Galanthus nivalis var. *maximus*	**Galanthus elwesii**
Galanthus nivalis var. *minus*	**Galanthus nivalis**
Galanthus nivalis var. *montanus*	**Galanthus nivalis**
Galanthus nivalis var. *octobrensis*	**Galanthus reginae-olgae** subsp. **reginae-olgae**
Galanthus nivalis var. *praecox*	**Galanthus reginae-olgae** subsp. **reginae-olgae**
Galanthus nivalis var. *rachelae*	**Galanthus reginae-olgae** subsp. **reginae-olgae**
Galanthus nivalis var. *redoutei*	**Galanthus alpinus** var. **alpinus**
Galanthus nivalis var. *reginae-olgae*	**Galanthus reginae-olgae** subsp. **reginae-olgae**
Galanthus nivalis var. *scharlockii* ✥	**Galanthus nivalis**
Galanthus nivalis var. *shaylockii* [*sic*]	**Galanthus nivalis**
Galanthus nivalis var. *typicus*	**Galanthus nivalis**
Galanthus octobrensis	**Galanthus reginae-olgae** subsp. **reginae-olgae**
Galanthus olgae	**Galanthus reginae-olgae** subsp. **reginae-olgae**
Galanthus olgae reginae	**Galanthus reginae-olgae** subsp. **reginae-olgae**
Galanthus perryi	**Galanthus × allenii**
Galanthus peshmenii	
Galanthus platyphyllus	
Galanthus plicatus*	
Galanthus plicatus auct. non M.Bieb	**Galanthus transcaucasicus**
Galanthus plicatus sens. Guss.	**Galanthus nivalis**
Galanthus plicatus subsp. **byzantinus**	
Galanthus plicatus subsp. *gueneri*	**Galanthus plicatus** subsp. **plicatus**
Galanthus plicatus subsp. *karamanoghluensis*	**Galanthus plicatus** subsp. **plicatus**
Galanthus plicatus subsp. **plicatus**	
Galanthus plicatus subsp. *plicatus* var. *viridifolius*	**Galanthus plicatus** subsp. **plicatus**
Galanthus plicatus subsp. *subplicatus*	**Galanthus nivalis**
Galanthus plicatus subsp. *vardarii*	**Galanthus plicatus** subsp. **plicatus**
Galanthus plicatus var. *byzantinus*	**Galanthus plicatus** subsp. **byzantinus**
Galanthus plicatus var. *genuinus* forma *excelsior*	**Galanthus plicatus** subsp. **plicatus**

*For explanation see page 2, point 4
*Voir les explications page 8, point 4
*Para mayor explicación, véase la página 14, point 4

Part I: All Names / Tous les Noms / Todos los Nombres

ALL NAMES	ACCEPTED NAME
Galanthus plicatus var. *genuinus* forma *maximus*	**Galanthus plicatus** subsp. **plicatus**
Galanthus plicatus var. *genuinus* forma *typicus*	**Galanthus plicatus** subsp. **plicatus**
Galanthus praecox	**Galanthus reginae-olgae** subsp. reginae-olgae
Galanthus rachelae	**Galanthus reginae-olgae** subsp. reginae-olgae
Galanthus redoutei	**Galanthus alpinus** var. **alpinus**
Galanthus reflexus auct. non Herb. ex Lindl.	**Galanthus nivalis**
Galanthus reflexus Herb. ex Lindl.	**Galanthus gracilis**
Galanthus reginae-olgae*	
Galanthus reginae-olgae auct. non Orph.	**Galanthus peshmenii**
Galanthus reginae-olgae subsp. *corcyrensis*	**Galanthus reginae-olgae** subsp. reginae-olgae
Galanthus reginae-olgae subsp. **reginae-olgae**	
Galanthus reginae-olgae subsp. **vernalis**	
Galanthus rizehensis	
Galanthus schaoricus	**Galanthus alpinus** var. **alpinus**
Galanthus sharlockii	**Galanthus nivalis**
Galanthus shaylockii [*sic*] ⊕	**Galanthus nivalis**
Galanthus transcaucasicus	
Galanthus valentinae	**Galanthus krasnovii**
Galanthus woronowii	
Oporanthus colchiciflorus	**Sternbergia colchiciflora**
Oporanthus fischerianus	**Sternbergia fischerina**
Oporanthus luteus	**Sternbergia lutea**
Oporanthus luteus var. *angustifolia*	**Sternbergia lutea**
Oporanthus luteus var. *latifolia*	**Sternbergia lutea**
Sternbergia aetnensis	**Sternbergia colchiciflora**
Sternbergia alexandrae	**Sternbergia colchiciflora**
Sternbergia americana	**Haylockia pusilla†**
Sternbergia aurantiaca	**Sternbergia lutea**
Sternbergia candida	
Sternbergia caucasica	**Merendera** sp.†
Sternbergia citrina	**Sternbergia colchiciflora**
Sternbergia clusiana	
Sternbergia colchiciflora	
Sternbergia colchiciflora var. *aetnensis*	**Sternbergia colchiciflora**
Sternbergia colchiciflora var. *alexandrae*	**Sternbergia colchiciflora**
Sternbergia colchiciflora var. *dalmatica*	**Sternbergia colchiciflora**
Sternbergia dalmatica	**Sternbergia colchiciflora**
Sternbergia exigua	**Tapeinanthus humilis (Narcissus canavillesii)** †
Sternbergia exscapa	**Sternbergia colchiciflora**
Sternbergia fischeriana	
Sternbergia fischeriana (Herb.) Rupr.	**Sternbergia fischeriana**
Sternbergia fischeriana forma *hissarica*	**Sternbergia fischeriana**
Sternbergia fischeriana subsp. *hissarica*	**Sternbergia fischeriana**
Sternbergia grandiflora	**Sternbergia clusiana**
Sternbergia greuteriana	
Sternbergia latifolia	**Sternbergia clusiana**
Sternbergia lutea*	
Sternbergia lutea Ker Gawl. ex Schult. & Schult.f.	**Sternbergia lutea***
Sternbergia lutea Orph.	**Sternbergia colchiciflora**
Sternbergia lutea subsp. *sicula*	**Sternbergia sicula**
Sternbergia lutea var. *graeca*	**Sternbergia sicula**

*For explanation see page 2, point 4
*Voir les explications page 8, point 4
*Para mayor explicación, véase la página 14, point 4

ALL NAMES	ACCEPTED NAME
Sternbergia lutea var. *sicula*	**Sternbergia sicula**
Sternbergia macrantha	**Sternbergia clusiana**
Sternbergia pulchella	
Sternbergia schubertii	
Sternbergia sicula	
Sternbergia spaffordiana	**Sternbergia clusiana**
Sternbergia stipitata	**Sternbergia clusiana**
Sternbergia vernalis	**Sternbergia fischeriana**

† Bulbous plants no longer included in *Sternbergia* (non CITES)
† Plantes bulbeuses n'étant plus incluses dans *Sternbergia* (non-CITES)
† Plantas bulbosas que ya no figuran en *Sternbergia* (no CITES)

✤ Names also currently used for cultivars (i.e. not of wild origin), including: *Galanthus nivalis* 'Atkinsii', *Galanthus nivalis* 'Imperati', *Galanthus nivalis* 'Scharlockii'
✤ Noms également utilisés pour les cultivars (c'est-à-dire d'origine non sauvage), y compris: *Galanthus nivalis* 'Atkinsii', *Galanthus nivalis* 'Imperati', *Galanthus nivalis* 'Scharlockii'
✤ Nombres utilizados frecuentemente para cultivares (es decir, no son de origen silvestre), inclusive: *Galanthus nivalis* 'Atkinsii', *Galanthus nivalis* 'Imperati', *Galanthus nivalis* 'Scharlockii'

Cultivar names are enclosed in single quotes, e.g. *Galanthus nivalis* 'Flore Pleno'
Les noms de cultivars sont mis entre guillemets simples
Los nombres de los cultivares se adjuntan entre comillas

*For explanation see page 2, point 4
*Voir les explications page 8, point 4
*Para mayor explicación, véase la página 14, point 4

31

PART II: NAMES IN CURRENT USE
Ordered alphabetically on accepted names and including geographical distribution

Cyclamen, *Galanthus* and *Sternbergia*

DEUXIEME PARTIE: NOMS ACCEPTÉS D'USAGE COURANT
Par ordre alphabétique des noms acceptés et avec répartition géographique

Cyclamen, *Galanthus* et *Sternbergia*

PARTE II: NOMBRES ACEPTADOS UTILIZADOS NORMALMENTE
Presentados por orden alfabético: nombres aceptados y inclusive la distribución geográfica

Cyclamen, *Galanthus* y *Sternbergia*

CYCLAMEN NAMES IN CURRENT USE
CYCLAMEN NOMS ACTUELLEMENT EN USAGE
CYCLAMEN NOMBRES UTILIZADOS NORMALMENTE

Cyclamen africanum Boiss. & Reut.
Cyclamen algeriense Jord.
Cyclamen ambiguum O.Schwarz
Cyclamen commutatum O.Schwarz & Lepper
Cyclamen neapolitanum sens. Duby
Cyclamen numidicum Glasau
Cyclamen pachylobum Jord.
Cyclamen saldense Pomel
Cyclamen subrotundum Jord.
Cyclamen venustum Jord.

Distribution: Algeria (N) and Tunisia (NW).

Cyclamen × atkinsii T. Moore (?*Cyclamen coum* × *Cyclamen persicum*)

Distribution: Only known from cultivation / Connu seulement en culture / Conocidas únicamente en cultivo.

Cyclamen balearicum Willk.
Cyclamen hederifolium Aiton subsp. *balearicum* O.Schwarz
Cyclamen repandum sens. ex R.Knuth.
Cyclamen repandum sens. Texidor
Cyclamen repandum Sm. subsp. *balearicum* (Willk.) Malag.
Cyclamen repandum Sm. var. *stenopetalum* Loret.
Cyclamen vernum Lobel ex Cambess.

Distribution: France (S), Spain (Balearic Is.).

Cyclamen cilicium Boiss. & Heldr.
Cyclamen cilicicum [misspelling of / faute d'orthographe de / falta de ortografía de *Cyclamen cilicium*]

Distribution: Turkey (S).

Cyclamen cilicium Boiss. & Heldr. forma **album** E.Frank & Koenen

Distribution: Turkey (S).

Cyclamen cilicium Boiss. & Heldr. forma **cilicium**

Distribution: Turkey (S).

Cyclamen colchicum (Albov) Albov
Cyclamen europaeum sens. Albov
Cyclamen europaeum L. subsp. *ponticum* (Albov) O.Schwarz

Part II: Cyclamen

Cyclamen europaeum L. var. *colchicum* Albov
Cyclamen europaeum L. var. *ponticum* Albov
Cyclamen ponticum (Albov) Poped.
Cyclamen purpurascens Mill. subsp. *ponticum* (Albov) Grey-Wilson

Distribution: Georgia.

Cyclamen coum Mill.

Distribution: Armenia, Azerbaijan, Bulgaria (E), Georgia, Iran (Islamic Republic of), Israel (N), Lebanon, Russian Federation (the) (incl. Crimea), Syrian Arab Republic (the) (W), Turkey.

Cyclamen coum Mill. subsp. caucasicum (K.Koch) O.Schwarz
Cyclamen abchasicum (Medw. ex Kusn.) Kolak.
Cyclamen adzharicum Poped.
Cyclamen atkinsii hort.
Cyclamen calcareum Kolak.
Cyclamen caucasicum Willd. ex Boiss.
Cyclamen caucasicum Willd. ex Steven
Cyclamen circassicum Poped.
Cyclamen coum Mill. var. *abchasicum* Medw. ex Kusn.
Cyclamen coum Mill. var. *caucasicum* (K.Koch) Meikle
Cyclamen coum Mill. var. *ibericum* Boiss.
Cyclamen europaeum L. var. *caucasicum* K.Koch
Cyclamen ibericum Goldie ex G.Don
Cyclamen ibericum Lem.
Cyclamen ibericum Steven ex Boiss.
Cyclamen ibericum T.Moore
Cyclamen vernum Sweet
Cyclamen vernum Sweet var. *caucasicum* O.Schwarz
Cyclamen zonale Jord.

Distribution: Armenia, Azerbaijan, Georgia, Russian Federation (the) (S), Turkey (NE).

Cyclamen coum Mill. subsp. coum forma albissimum R.H.Bailey, Koenen, Lillywh. & P.J.M.Moore

Distribution: Turkey.

Cyclamen coum Mill. subsp. coum forma coum
Cyclamen apiculatum Jord.
Cyclamen atkinsii Glasau
Cyclamen brevifrons Jord.
Cyclamen coum Mill. subsp. *hiemale* (Hildebr.) O.Schwarz
Cyclamen durostoricum Pantu & Solacolu
Cyclamen europaeum Pall.
Cyclamen hiemale Hildebr.
Cyclamen hyemale Salisb.
Cyclamen kusnetzovii Kotov & Czernova
Cyclamen orbiculatum Mill.
Cyclamen orbiculatum Mill. var. coum (Mill.) Door.
Cyclamen vernale hort.

Cyclamen vernum Sweet var. *hiemale* (Hildebr.) O.Schwarz forma *pseudocoum* O.Scharwz
Cyclaminus coa (Mill.) Asch.

Distribution: Bulgaria (E), Israel (N), Lebanon, Russian Federation (the) (incl. Crimea), Syrian
  Arab Republic (the) (W), Turkey (N).

Cyclamen coum Mill. subsp. **coum** forma **pallidum** Grey-Wilson

Distribution: Turkey (N).

Cyclamen coum Mill. subsp. **elegans** (Boiss. & Buhse) Grey-Wilson
Cyclamen elegans Boiss. & Buhse

Distribution: Iran (Islamic Republic of) (N).

Cyclamen creticum (Dörfl.) Hildebr.

Distribution: Greece (Crete, Karpathos).

Cyclamen creticum (Dörfl.) Hildebr. forma **creticum**
Cyclamen hederifolium Aiton subsp. *creticum* (Dörfl.) O.Schwarz
Cyclamen repandum Sm. var. *creticum* Dörfl.

Distribution: Greece (Crete, Karpathos).

Cyclamen creticum (Dörfl.) Hildebr. forma **pallide-roseum** Grey-Wilson

Distribution: Greece (Crete, Karpathos).

Cyclamen cyprium Kotschy
Cyclamen neapolitanum sens. Boiss.

Distribution: Cyprus.

Cyclamen × drydeniae Grey-Wilson (*Cyclamen coum × Cyclamen trochopteranthum*)

Distribution: Only known from cultivation / Connu seulement en culture / Conocidas únicamente
  en cultivo.

Cyclamen graecum Link

Distribution: Cyprus, Greece, Turkey.

Cyclamen graecum Link subsp. **anatolicum** Ietsw.
?*Cyclamen cyprium* Glasau

Part II: Cyclamen

Cyclamen cypro-graecum E.Mutch & N.Mutch
Cyclamen maritimum Hildebr.
Cyclamen pseudomaritimum Hildebr.

Distribution: Cyprus (N), Greece (E. Aegean Is., incl. Rhodes), Turkey (S and SW).

Cyclamen graecum Link subsp. **graecum** forma **album** R.Frank

Distribution: Greece (Peleponnese).

Cyclamen graecum Link subsp. **graecum** forma **graecum**
Cyclamen aegineticum Hildebr.
Cyclamen gaidurowryssii Glasau
Cyclamen gaydurowryssii (orthographical variant) Glasau
Cyclamen miliarakesii Heldr. ex Halácsy
Cyclamen miliarakesii Heldr. ex Hildebr.
Cyclamen pentelici Hildebr.
Cyclamen persicum sens. Sibth. & Sm.
Cyclamen velutinum Jord.
Cyclaminos graeca (Link) Heldr.
Cyclaminos miliarakesii Heldr.
Cyclaminus graeca (Link) Asch.

Distribution: Greece (incl. Crete, Aegean Is.).

Cyclamen graecum Link subsp. **mindleri** (Heldr.) A.P.Davis & Govaerts, (see page 129).
Cyclamen graecum Link. subsp. *candicum* Ietsw.
Cyclamen persicum subsp. *mindleri* (Heldr.) Knuth
Cyclamen pseudograecum Hildebr.
Cyclaminos mindleri Heldr.

Distribution: Greece (W. Crete).

Cyclamen hederifolium Aiton

Distribution: Albania, Bulgaria, France (incl. Corsica), Greece (incl. Crete), Italy (incl. Sardinia,
  Sicily), Switzerland, Turkey (W), Yugoslavia (former).
Introduced: United Kingdom of Great Britain and Northern Ireland (an escape from gardens / une
  évasion des jardins / un escape de jardines).

Cyclamen hederifolium Aiton var. **confusum** Grey-Wilson

Distribution: Greece (S. Greece, Aegean Is., Crete), Italy (Sicily).

Cyclamen hederifolium Aiton var. **hederifolium** forma **albiflorum** (Jord.) Grey-Wilson
Cyclamen albiflorum Jord.

Distribution: Albania, Bulgaria, France, Greece (incl. Crete), Italy (incl. Sicily), Switzerland,
  Turkey, Yugoslavia (former).

Cyclamen hederifolium Aiton var. **hederifolium** forma **hederifolium**
 Cyclamen aedirhizum Jord.
 Cyclamen angulare Jord.
 Cyclamen cyclaminus Bedevian
 Cyclamen europaeum L., pro parte
 Cyclamen hederifolium Aiton subsp. *romanum* (Griseb.) O.Schwarz
 Cyclamen hederifolium Sibth. & Sm.
 Cyclamen insulare Jord.
 Cyclamen linaerifolium DC.
 Cyclamen neapolitanum Ten.
 Cyclamen poli Chiaje
 Cyclamen romanum Griseb.
 Cyclamen sabaudum Jord.
 Cyclamen subhastatum Rchb.
 Cyclaminus neopolitana (Ten.) Asch.

Distribution: Albania, Bulgaria, France, Greece, Italy, Switzerland, Turkey, Yugoslavia (former).
Introduced: United Kingdom of Great Britain and Northern Ireland (an escape from gardens / une évasion des jardins / un escape de jardines).

Cyclamen × **hildebrandii** O.Schwarz (*Cyclamen africanum* × *Cyclamen hederifolium*)

Distribution: Only known from cultivation / Connu seulement en culture / Conocidas únicamente en cultivo.

Cyclamen intaminatum (Meikle) Grey-Wilson
 Cyclamen cilicium var. [*sic*] Turrill
 Cyclamen cilicium Boiss. & Heldr. var. *alpinum* hort.
 Cyclamen cilicium Boiss. & Heldr. var. *intaminatum* Meikle

Distribution: Turkey (W. and SW).

Cyclamen libanoticum Hildebr.

Distribution: Lebanon.

Cyclamen × **meiklei** Grey-Wilson (*Cyclamen creticum* × *Cyclamen repandum*)

Distribution: Only known from cultivation / Connu seulement en culture / Conocidas únicamente en cultivo.

Cyclamen mirabile Hildebr.

Distribution: Turkey (SW).

Cyclamen mirabile Hildebr. forma **mirabile**

Part II: Cyclamen

Distribution: Turkey (SW).

Cyclamen mirabile Hildebr. forma **niveum** J.White & Grey-Wilson

Distribution: Turkey (SW).

Cyclamen parviflorum Poped.

Distribution: Turkey (NE).

Cyclamen parviflorum Poped. var. **parviflorum**

Distribution: Turkey (NE).

Cyclamen parviflorum Poped. var. **subalpinum** Grey-Wilson

Distribution: Turkey (NE).

Cyclamen persicum Mill.

Distribution: Algeria, Cyprus, Greece (E. Greece, Crete (E), Karpathos, E. Aegean Is., Rhodes), Israel, Jordan, Lebanon, Syrian Arab Republic (the) (W), Tunisia (N), Turkey (S).

Cyclamen persicum Mill. var. **autumnale** Grey-Wilson

Distribution: Israel (N).

Cyclamen persicum Mill. var. **persicum** forma **albidum** (Jord.) Grey-Wilson
  Cyclamen albidum Jord.

Distribution: Algeria, Cyprus, Greece (Crete (E), Karpathos, E. Aegean Is., Rhodes), Israel, Jordan, Lebanon, Syrian Arab Republic (the), Tunisia, Turkey.

Cyclamen persicum Mill. var. **persicum** forma **persicum**
  Cyclamen aleppicum Fisch. ex Hoffmanns.
  Cyclamen antilochium Decne
  Cyclamen cyprium Sibth.
  Cyclamen hederaceum Sieber ex Steud.
  Cyclamen hederifolium Kotschy
  Cyclamen latifolium Sm.
  Cyclamen persicum Mill. subsp. *eupersicum* Knuth
  Cyclamen punicum Pomel
  Cyclamen pyrolifolium Salsib.
  Cyclamen tunetanum Jord.
  Cyclamen utopicum Hoffmanns.
  Cyclamen vernale Mill.

Cyclaminus persica (Mill.) Asch.

Distribution: Algeria, Cyprus, Greece (Crete (E), Karpathos, E. Aegean Is., Rhodes), Israel, Jordan, Lebanon, Syrian Arab Republic (the), Tunisia, Turkey (S & SW).

Cyclamen persicum Mill. var. **persicum** forma **puniceum** (Glasau) Grey-Wilson
Cyclamen aleppicum Hoffmanns. subsp. *puniceum* Glasau

Distribution: Found in the eastern end of species range only, precise distribution unknown / Présent uniquement dans la partie orientale de l'aire de répartition de l'espèce / Probablemente a lo largo del área de distribución de la especie:
?Jordan, ?Lebanon, Syrian Arab Republic (the).

Cyclamen persicum Mill. var. **persicum** forma **roseum** Grey-Wilson nom. provis.

Distribution: Found in the eastern end of species range only, precise distribution unknown / Présent uniquement dans la partie orientale de l'aire de répartition de l'espèce / Probablemente a lo largo del área de distribución de la especie:
?Jordan, ?Lebanon, Syrian Arab Republic (the).

Cyclamen pseudibericum Hildebr.

Distribution: Turkey (S).

Cyclamen pseudibericum Hildebr. forma **pseudibericum**
Cyclamen libanoticum Hildebr. subsp. *pseudibericum* (Hildebr.) Glasau

Distribution: Turkey (SE).

Cyclamen pseudibericum Hildebr. forma **roseum** Grey-Wilson

Distribution: Turkey (SE).

Cyclamen purpurascens Mill.

Distribution: Austria, Czech Republic (the), France (E), Germany (S), Hungary, Italy (N), Poland (S), Slovakia, Switzerland, Yugoslavia (former).
Introduced: Romania (naturalized / naturalisée / naturalizada), Russian Federation (the) (naturalized / naturalisée / naturalizada).

Cyclamen purpurascens Mill. forma **album** Grey-Wilson

Distribution: Probably throughout the range of the species / Dans tout probablement l'intervalle des espèces / Probablemente a través del rango de la especie: Austria, Czech Republic (the), France (E), Germany (S), Hungary, Italy (N), Poland (S), Slovakia, Switzerland, Yugoslavia (former).

41

Part II: Cyclamen

Cyclamen purpurascens Mill. forma **purpurascens**
Cyclamen aestivum Rchb.
Cyclamen breviflorum Jord.
Cyclamen clusii Lindl.
Cyclamen cyclophyllum Jord.
Cyclamen deltoideum Tausch
Cyclamen europaeum L., pro parte
Cyclamen europaeum L. subsp. *orbiculatum* (Mill.) O.Schwarz
Cyclamen europaeum L. subsp. *orbiculatum* (Mill.) O.Schwarz var. *immaculatum* Hrabetova
Cyclamen europaeum L. var. *typicum* Albov
Cyclamen europaeum Savi
Cyclamen europaeum sens. Aiton
Cyclamen fatrense Halda & Soják
Cyclamen floridum Salisb.
Cyclamen hastatum Tausch
Cyclamen holochlorum Jord.
Cyclamen lilacinum Jord.
Cyclamen littorale Sadler ex Rchb.
Cyclamen purpurascens Mill. subsp. *immaculatum* (Hrabetova) Halda & Soják
Cyclamen retroflexum Moench
Cyclamen umbratile Jord.
Cyclamen variegatum Pohl

Distribution: Austria, Czech Republic (the), France, Germany, Hungary, Italy (N), Poland,
   Slovakia, Switzerland, Yugoslavia (former).
Introduced: Romania (naturalized / naturalisée / naturalizada), Russian Federation (the)
(naturalized / naturalisée / naturalizada).

Cyclamen repandum Sm.

Distribution: France (SE, incl. Corsica), Greece, Italy (incl. Sardinia), Switzerland (S), Yugoslavia
   (former).

Cyclamen repandum Sm. subsp. **peloponnesiacum** var. **peloponnesiacum** (Grey-Wilson)
Grey-Wilson
 Cyclamen repandum Sm. subsp. *peloponnesiacum* Grey-Wilson forma *peloponnesiacum* Grey-
Wilson

Distribution: Greece (central Peloponnese).

Cyclamen repandum Sm. subsp. **peloponnesiacum** Grey-Wilson var. **vividum** (Grey-
Wilson) Grey-Wilson
Cyclamen eucardium Jord.
Cyclamen repandum Sm. subsp. *peloponnesiacum* Grey Wilson forma *vividum* Grey-Wilson
Cyclamen spectabile Jord.
Cyclamen stenopetalum Jord.

Distribution: Greece (E. Peloponnese).

Cyclamen repandum Sm. subsp. **repandum** var. **baborense** Debussche & Quézel
Cyclamen repandum Sm. subsp. *atlanticum* Maire

Cyclamen baborense Batt.

Distribution: Algeria.

Cyclamen repandum Sm. subsp. **repandum** var. **repandum** forma **album** Grey-Wilson

Distribution: France, Greece, Italy, ?Switzerland (S), Yugoslavia (former).

Cyclamen repandum Sm. subsp. **repandum** var. **repandum** forma **repandum**
Cyclamen europeaum L., pro parte
Cyclamen ficariifolium Rchb.
Cyclamen hederifolium Sims
Cyclamen ilicetorum Jord.
Cyclamen lobospilum Jord.
Cyclamen rarinaevum Jord.
Cyclamen vernale sens. O.Schwarz
Cyclamen vernum Lobel ex Rchb.
Cyclaminus repanda (Sm.) Asch.

Distribution: France (SE, incl. Corsica), Italy (incl. Sardinia), Switzerland (S), Yugoslavia (former).

Cyclamen repandum Sm. subsp. **rhodense** (Miekle) Grey-Wilson
Cyclamen repandum Sm. var. *rhodense* Meikle
Cyclamen rhodium R.Gorer ex O.Schwarz & Lepper

Distribution: Greece (Rhodes, Cos).

Cyclamen rohlfsianum Asch.

Distribution: Libyan Arab Jamahiriya (the) (N).

Cyclamen × saundersiae Grey-Wilson (*Cyclamen balearicum* × *Cyclamen repandum*)

Distribution: Only known from cultivation / Connu seulement en culture / Conocidas únicamente en cultivo.

Cyclamen × schwarzii Grey-Wilson (*Cyclamen libanoticum* × *Cyclamen pseudibericum*)

Distribution: Only known from cultivation / Connu seulement en culture / Conocidas únicamente en cultivo.

Cyclamen somalense Thulin & Warfa

Distribution: Somalia (N).

Part II: Cyclamen

Cyclamen trochopteranthum O.Schwarz

Distribution: Turkey (SW).

Cyclamen trochopteranthum O.Schwarz forma **leucanthum** Grey-Wilson

Distribution: Turkey (SW).

Cyclamen trochopteranthum O.Schwarz forma **trochopteranthum**
  Cyclamen alpinum sens. Turrill, pro parte
  Cyclamen orbiculatum Mill. var. *alpinum* Saunders

Distribution: Turkey (SW).

Cyclamen × **wellensiekii** Ietsw. (*Cyclamen cyprium* × *Cyclamen libanoticum*)

Distribution: Only known from cultivation / Connu seulement en culture / Conocidas únicamente
  en cultivo.

Cyclamen × **whiteae** Grey-Wilson (*Cyclamen graecum* × *Cyclamen hederifolium*)

Distribution: Only known from cultivation / Connu seulement en culture / Conocidas únicamente
  en cultivo.

Names of uncertain affinity / Noms d'affinité incertaine / Nombres ambiguos

Cyclamen autumnale Boos
Cyclamen cilicium Hildebr.
Cyclamen cordifolium Stokes
Cyclamen coum sens. Rchb.
Cyclamen crassifolium Hildebr.
Cyclamen europaeum Sm.
Cyclamen hederifolium Willd.
Cyclamen immaculatum Pieri
Cyclamen intermedium Wender.
Cyclamen jovis Hildebr.
Cyclamen macrophyllum Sieber
Cyclamen macropus Zucc.
Cyclamen × *marxii* Halda
Cyclamen officinale Wender.
Cyclamen rotundifolium St.-Lég.
Cyclamen tauricum hort. Dammann ex Sprenger
Cyclaminum vernum Bubani

Ambiguous names / Noms ambigus / Nombres ambiguos

Cyclamen europaeum L. (includes **C. hederifolium, C. purpurascens, C. repandum**)
Cyclaminus europaea (L.) Asch.

Cyclaminus europaeus [*sic*] (L.) Scop.

Names causing confusion / Noms source de confusion / Nombres que ocasionan confusión

Cyclamen alpinum hort. Dammann ex Sprenger
Cyclamen alpinum hort. Dammann ex Sprenger var. *album* hort. Dammann ex Sprenger
Cyclamen coum Mill. subsp. *alpinum* (hort. Dammann ex Sprenger) O.Schwarz
Cyclamen europaeum sens. Mill.
Cyclamen indicum L.
Cyclamen vernum Sweet forma *alpinum* (hort. Dammann ex Sprenger) O.Schwarz
Cyclamen vernum Sweet var. *hiemale* (Hildebr.) O.Schwarz forma *alpinum* (hort. Dammann ex Sprenger) O.Schwarz

GALANTHUS NAMES IN CURRENT USE
GALANTHUS NOMS ACTUELLEMENT EN USAGE
GALANTHUS NOMBRES UTILIZADOS NORMALMENTE

Galanthus × **allenii** Baker (?*Galanthus alpinus* × *Galanthus woronowii* Losinsk.)
Galanthus allenii Baker
Galanthus latifolius Rupr. forma *allenii* (Baker) Beck
Galanthus nivalis L. subsp. *allenii* (Baker) Gottl.-Tann.
Galanthus perryi hort. Ware ex Baker

Distribution: Origin unknown, probably only known from cultivation / connu seulement en culture
/ conocidas únicamente en cultivo.

Galanthus alpinus Sosn.

Distribution: Armenia, Azerbaijan, Georgia, ?Iran (Islamic Republic of), Russian Federation (the),
Turkey.
Introduced: United Kingdom of Great Britain and Northern Ireland.

Galanthus alpinus Sosn. var. **alpinus**
Galanthus caucasicus (Baker) Burb.
Galanthus caucasicus (Baker) Grossh.
Galanthus cilicicus Baker subsp. *caucasicus* O.Schwarz
Galanthus grandis Burb.
Galanthus nivalis sens. Ledeb.
Galanthus nivalis L. subsp. *caucasicus* Baker
Galanthus nivalis L. var. *caucasicus* (Baker) Beck
Galanthus nivalis L. var. *caucasicus* (Baker) Fomin
Galanthus nivalis L. var. *caucasicus* (Baker) J. Phillippow
Galanthus nivalis L. var. *major* Redouté ex Rupr.
Galanthus nivalis L. var. *major* sens. Fiori, non Redouté
Galanthus nivalis L. var. *redoutei* Rupr. ex Regel
Galanthus redoutei (Rupr. ex Regel) Regel
Galanthus schaoricus Kem.-Nath.

Distribution: Armenia, Azerbaijan, Georgia, ?Iran (Islamic Republic of), Russian Federation (the),
Turkey.

Galanthus alpinus Sosn. var. **bortkewitschianus** (Koss) A.P.Davis
Galanthus bortkewitschianus Koss

Distribution: Russian Federation (the) (Caucasus).

Galanthus angustifolius Koss
Galanthus nivalis L. subsp. *angustifolius* (Koss) Artjush.

Distribution: Russian Federation (the) (Caucasus).

Part II: Galanthus

Galanthus cilicicus Baker
Galanthus nivalis L. subsp. *cilicicus* (Baker) Gottl.-Tann.

Distribution: Turkey (S).

Galanthus elwesii Hook.f., nom. cons.
Chianthemum elwesii (Hook.f.) Kuntze
Chianthemum graecum (Orph. ex Boiss.) Kuntze
Galanthus bulgaricus Velen.
Galanthus caucasicus (Baker) Grossh. var. *hiemalis* Stern
Galanthus caucasicus hort.
Galanthus elwesii Hook.f. subsp. *akmanii* N.Zeybek
Galanthus elwesii Hook.f. subsp. *baytopii* (N.Zeybek) N.Zeybek & E.Sauer
Galanthus elwesii Hook.f. subsp. *melihae* N.Zeybek
Galanthus elwesii Hook.f. subsp. *tuebitaki* N.Zeybek
Galanthus elwesii Hook.f. subsp. *wagenitzii* N.Zeybek
Galanthus elwesii Hook.f. subsp. *yayintaschii* N.Zeybek
Galanthus elwesii Hook.f. [var.] *globosus* Ewbank
Galanthus elwesii Hook.f. var. *maximus* (Velen.) Beck
Galanthus elwesii Hook.f. var. *monostictus* P.D.Sell
Galanthus elwesii Hook.f. var. *platyphyllus* Kamari
Galanthus elwesii Hook.f. var. *robustus* Baker
Galanthus elwesii Hook.f. var. *whittallii* (hort.) W.Irving
Galanthus elwesii Hook.f. var. *whittallii* Moon
Galanthus elwesii Hook.f. var. *whittallii* S.Arn.
Galanthus globosus Burb.
Galanthus globosus Wilks
Galanthus gracilis Čelak. subsp. *baytopii* N.Zeybek
Galanthus graecus Orph. ex Boiss.
Galanthus graceus Orph. ex Boiss. forma *maximus* (Velen.) Zahar.
Galanthus graecus Orph. ex Boiss. var. *maximus* (Velen.) Hayek
Galanthus maximus Velen.
Galanthus melihae (N.Zeybek) N.Zeybek & E.Sauer
Galanthus nivalis L. subsp. *elwesii* (Hook.f.) Gottl.-Tann.
Galanthus nivalis L. subsp. *graecus* (Orph. ex Boiss.) Gottl.-Tann., pro parte excl. *G. gracilis* Čelak
Galanthus nivalis L. var. *maximus* (Velen.) Stoj. & Stevanov

Distribution: Bulgaria, Greece, Turkey, Ukraine (S), Yugoslavia (former).
Introduced: The Netherlands (very rare, an occasional escape / très rare, occasionnellement échappé / Muy rara, probablemente se volvió silvestre), United Kingdom of Great Britain and Northern Ireland (rare, an occasional escape / rare, occasionnellement échappé / rara, probablemente se volvió silvestre).

Galanthus fosteri Baker
Galanthus fosteri Baker var. *antepensis* N.Zeybek & E.Sauer
Galanthus latifolius Rupr. forma *fosteri* (Baker) Beck

Distribution: ?Israel, Jordan, Lebanon, Syrian Arab Republic (the), Turkey.

Galanthus gracilis Čelak.
Galanthus elwesii Hook.f. subsp. *minor* D.A.Webb
Galanthus elwesii Hook.f. var. *reflexus* (Herb. ex Lindl.) Beck, nom. rej. prop.
Galanthus elwesii Hook.f. var. *stenophyllus* Kamari

48

Galanthus graecus auct. non Orph. ex Boiss.
Galanthus graceus Orph. ex Boiss. forma *gracilis* (Čelak.) Zahar.
Galanthus reflexus Herb. ex Lindl. nom. rej. prop.

Distribution: Bulgaria, Greece, Romania, Turkey, Ukraine (S).

Galanthus × grandiflorus Baker
Galanthus × maximus Baker (*Galanthus plicatus* × *Galanthus nivalis*)

Distribution: Only known from cultivation / Connu seulement en culture / Conocidas únicamente
en cultivo. Garden Hybrid / Hybride de jardin / Híbrido de jardín.

Galanthus ikariae Baker
Galanthus ikariae Baker subsp. *snogerupii* Kamari

Distribution: Greece (Aegean Is.).

Galanthus koenenianus Lobin, C.D.Brickell & A.P.Davis

Distribution: Turkey (NE).

Galanthus krasnovii A.P.Khokhr.
Galanthus krasnovii A.P.Khokhr. subsp. *maculatus* A.P.Khokhr.
Galanthus valentinae Panjut. ex Grossh.

Distribution: Georgia (W), Turkey (NE).

Galanthus lagodechianus Kem.-Nath.
Galanthus cabardensis Koss
Galanthus kemulariae Kuth.
Galanthus ketzkhovelii Kem.-Nath.

Distribution: Armenia, Azerbaijan, Georgia, Russian Federation (the) (Caucasus).

Galanthus nivalis L. (see Annex IV for further synonyms)
Chianthemum nivale (L.) Kuntze
Galanthus alexandrii Porcius
Galanthus atkinsii hort. Barr
Galanthus imperati Bertol.
Galanthus montana Schur
Galanthus nivalis L. forma *pictus* Maly
Galanthus nivalis L. subsp. *humboldtii* N.Zeybek
Galanthus nivalis L. subsp. *imperati* (Bertol.) Baker
Galanthus nivalis L. subsp. *subplicatus* (N.Zeybek) N.Zeybek & E.Sauer
Galanthus nivalis L. [var.] *atkinsii* J.Allen
Galanthus nivalis L. var. *atkinsii* Mallett
Galanthus nivalis L. var. *carpaticus* Fodor
Galanthus nivalis L. var. *europaeus* Beck forma *hololeucus* (Čelak.) Beck
Galanthus nivalis L. var. *europaeus* Beck forma *hortensis* (Herb.) Beck
Galanthus nivalis L. var. *europaeus* Beck forma *scharloki* [*sic*] (Casp.) Beck
Galanthus nivalis L. var. *grandior* Schult. & Schult.f.

Part II: Galanthus

Galanthus nivalis L. var. *hololeucus* Čelak.
Galanthus nivalis L. var. *hortensis* Herb.
Galanthus nivalis L. var. *imperati* Mallett
Galanthus nivalis L. var. *major* sens. Fiori
Galanthus nivalis L. var. *majus* [*sic*] Ten.
Galanthus nivalis L. var. *minus* Ten.
Galanthus nivalis L. var. *montanus* (Schur) Rouy
Galanthus nivalis L. var. *scharlockii* Casp.
Galanthus nivalis L. var. *shaylockii* [*sic*] Harpur-Crewe
Galanthus nivalis L. var. *typicus* Rouy
Galanthus plicatus sens. Guss.
Galanthus plicatus M.Bieb. subsp. *subplicatus* N.Zeybek
Galanthus reflexus auct. non Herb. ex Lindl.
Galanthus sharlockii (Casp.) Baker
Galanthus shaylockii [*sic*] J.Allen

Distribution: Albania, Austria, Bulgaria, Czech Republic (the), France, Germany, Greece, Hungary, Italy, Poland, Republic of Moldova (the), Romania, Slovakia, ?Spain, Switzerland, Turkey (European Turkey), Ukraine, Yugoslavia (former).
Introduced: Belgium, Denmark (rare), The Netherlands, Sweden (rare), United Kingdom of Great Britain and Northern Ireland (locally frequent / localement fréquents / localmente frecuentes), and certain places in / et certains en / y ciertos en France (N) and Germany.

Galanthus nivalis L. 'Flore Pleno'
Galanthus nivalis L. forma *pleniflorus* P.D.Sell

Distribution: Generally only known from cultivation / Généralement connu en culture / Generalmente conocidas sólo en cultivo.

Galanthus peshmenii A.P.Davis & C.D.Brickell
Galanthus cilicicus auct. non Baker
Galanthus nivalis L. subsp. *cilicicus* auct. non (Baker) Gottl.-Tann.
Galanthus reginae-olgae auct. non Orph.

Distribution: Greece (E. Aegean Is.), Turkey (S).

Galanthus platyphyllus Traub & Moldenke
Galanthus ikariae Baker subsp. *latifolius* Stern, pro parte
Galanthus latifolius Rupr.
Galanthus latifolius Rupr. forma *typicus* Beck
Galanthus latifolius Rupr. forma *typicus* Gottl.-Tann.

Distribution: Georgia, Russian Federation (the) (Caucasus).

Galanthus plicatus M.Bieb.

Distribution: Romania, Russian Federation (the) (incl. Crimea), Turkey.
Introduced: United Kingdom of Great Britain and Northern Ireland (rare, an occasional escape / rare, occasionnellement échappé / rara, probablemente se volvió silvestre).

Galanthus plicatus M.Bieb. subsp. **byzantinus** (Baker) D.A.Webb
Galanthus byzantinus Baker
Galanthus byzantinus Baker subsp. *brauneri* N.Zeybek
Galanthus byzantinus Baker subsp. *saueri* N.Zeybek
Galanthus byzantinus Baker subsp. *tughrulii* N.Zeybek
Galanthus nivalis L. subsp. *byzantinus* (Baker) Gottl.-Tann.
Galanthus plicatus M.Bieb. var. *byzantinus* (Baker) Beck

Distribution: Turkey (NW).

Galanthus plicatus M.Bieb. subsp. **plicatus**
Chianthemum plicatum (M.Bieb.) Kuntze
Galanthus clusii Fisch. sens. Steud.
Galanthus latifolius Salisb.
Galanthus nivalis L. subsp. *plicatus* (M.Bieb.) Gottl.-Tann.
Galanthus plicatus M.Bieb. subsp. *gueneri* N.Zeybek
Galanthus plicatus M.Bieb. subsp. *karamanoghluensis* N.Zeybek
Galanthus plicatus M.Bieb. subsp. *plicatus* var. *virdifolius* P.D.Sell
Galanthus plicatus M.Bieb. subsp. *vardarii* N.Zeybek
Galanthus plicatus M.Bieb. var. *genuinus* forma *excelsior* Beck
Galanthus plicatus M.Bieb. var. *genuinus* forma *maximus* Beck
Galanthus plicatus M.Bieb. var. *genuinus* forma *typicus* Beck

Distribution: Romania, Russian Federation (the) (Crimea), Turkey.
Introduced: United Kingdom of Great Britain and Northern Ireland (rare, an occasional escape /
rare, occasionnellement échappé / rara, probablemente se volvió silvestre).

Galanthus reginae-olgae Orph.

Distribution: Greece, Italy (Sicily), Yugoslavia (former).

Galanthus reginae-olgae Orph. subsp. **reginae-olgae**
Chianthemum olgae (Orph.) Kuntze
Galanthus corcynensis [*sic*] T.Shortt
Galanthus corcyrensis (Beck) Stern
Galanthus corcyrensis Burb.
Galanthus corcyrensis J.Allen
Galanthus corcyrensis Leichtlin ex Correvon
Galanthus corcyrensis (*praecox*) [*sic*] hort. ex Baker
Galanthus elsae Burb.
Galanthus elsae Ewbank
Galanthus elsae J.Allen
Galanthus imperati Bertol. forma *australis* Zodda
Galanthus nivalis L. forma *octobrinus* hort. ex Voss
Galanthus nivalis L. subsp. *reginae-olgae* (Orph.) Gottl.-Tann.
Galanthus nivalis L. var. *corcyrensis* (Beck) Halácsy
Galanthus nivalis L. [var.] *corcyrensis* hort. ex Leichtlin
Galanthus [*nivalis* L.] var. *elsae* Mallett
Galanthus nivalis L. var. *europaeus* Beck forma *corcyrensis* hort. ex Beck
Galanthus nivalis L. var. *europaeus* Beck forma *olgae* (Orph.) Beck
Galanthus nivalis L. var. *octobrensis* Mallett
Galanthus nivalis L. var. *praecox* Mallctt
Galanthus nivalis L. var. *rachelae* Mallett
Galanthus nivalis L. var. *reginae-olgae* (Orph.) Fiori

Part II: Galanthus

Galanthus octobrensis Burb.
Galanthus octobrensis Ewbank
Galanthus octobrensis hort. ex Baker
Galanthus octrobrensis hort. ex Burb.
Galanthus octobrensis J.Allen
Galanthus octobrensis Leichtlin ex Correvon
Galanthus octobrensis T.Shortt
Galanthus olgae Orph. ex Boiss.
Galanthus olgae reginae hort. ex Leichtlin
Galanthus praecox Burb.
Galanthus praecox J.Allen
Galanthus rachelae Burb.
Galanthus rachelae Ewbank
Galanthus rachelae J.Allen
Galanthus reginae-olgae Orph. subsp. *corcyrensis* (Beck) Kamari

Distribution: Greece (mainland /continent / continente and Corfu), Italy (Sicily).

Galanthus reginae-olgae Orph. subsp. **vernalis** Kamari

Distribution: ?Albania, Greece, Italy (Sicily), Yugoslavia (former).

Galanthus rizehensis Stern
Galanthus cilicicus auct. non Baker
Galanthus glaucescens A.P.Khokhr.
Galanthus latifolius Rupr. [var.] *rizaensis* [*sic*] Anon.
Galanthus latifolius Rupr. var. *rizehensis* Stern & Gilmour

Distribution: Georgia, Russian Federation (the) (Black Sea Coast), Turkey (N).

Galanthus transcaucasicus Fomin
Galanthus caspius (Rupr.) Grossh.
Galanthus nivalis L. var. *caspius* Rupr.
Galanthus plicatus auct. non M.Bieb.

Distribution: Armenia, Azerbaijan, Iran (Islamic Republic of) (N).

Galanthus woronowii Losinsk.
Galanthus ikariae auct. non Baker, pro parte
Galanthus ikariae Baker subsp. *latifolius* Stern pro parte
Galanthus latifolius auct. non Rupr.

Distribution: Georgia, Russian Federation (the) (Black Sea Coast area), Turkey (NE).
Introduced: The Netherlands (rare, an occasional escape / rare, occasionnellement échappé / rara, probablemente se volvió silvestre), United Kingdom of Great Britain and Northern Ireland (rare, an occasional escape / rare, occasionnellement échappé / rara, probablemente se volvió silvestre).

STERNBERGIA NAMES IN CURRENT USE
STERNBERGIA NOMS ACTUELLEMENT EN USAGE
STERNBERGIA NOMBRES UTILIZADOS NORMALMENTE

Sternbergia candida B.Mathew & T.Baytop

Distribution: Turkey (SW).

Sternbergia clusiana (Ker Gawl.) Ker Gawl. ex Spreng.
  Amaryllis clusiana Ker Gawl.
  Sternbergia grandiflora Boiss. ex Baker
  Sternbergia latifolia Boiss. & Hausskn. ex Baker
  Sternbergia macrantha J.Gay ex Baker
  Sternbergia spaffordiana Dinsm.
  Sternbergia stipitata Boiss. & Hausskn. ex Boiss.

Distribution: Greece (Samos), Iran (Islamic Republic of), Iraq, Israel, Jordan, Lebanon, Syrian
  Arab Republic (the), Turkey.

Sternbergia colchiciflora Waldst. & Kit.
  Amaryllis aetnensis Raf.
  Amaryllis citrina Sibth. & Sm.
  Amaryllis colchiciflora (Waldst. & Kit.) Ker Gawl.
  Oporanthus colchiciflorus (Waldst. & Kit.) Herb.
  Sternbergia aetnensis (Raf.) Guss.
  Sternbergia alexandrae Sosn.
  Sternbergia citrina (Sibth. & Sm.) Schult. & Schult.f.
  Sternbergia colchiciflora Waldst. & Kit. var. *aetnensis* (Raf.) Rouy
  Sternbergia colchiciflora Waldst. & Kit. var. *alexandrae* (Sosn.) Artjush.
  Sternbergia colchiciflora Waldst. & Kit. var. *dalmatica* Herb.
  Sternbergia dalmatica (Herb.) Herb.
  Sternbergia exscapa Tineo ex Guss.
  Sternbergia lutea Orph.

Distribution: Armenia, Azerbaijan, Bulgaria, Georgia, Greece, Hungary, Iran (Islamic Republic
  of), Israel, Italy (Sicily), Romania, Russian Federation (the) (incl. Crimea), Spain (SE), Turkey,
  Yugoslavia (former).

Sternbergia fischeriana (Herb.) M.Roem.
  Amaryllis lutea M.Bieb.
  Amaryllis vernalis Mill., nom. rej. prop.
  Oporanthus fischerianus Herb.
  Sternbergia fischeriana (Herb.) M.Roem. forma *hissarica* Kapinos
  Sternbergia fischeriana (Herb.) M.Roem. subsp. *hissarica* (Kapinos) Artjush.
  Sternbergia fischeriana (Herb.) Rupr.
  Sternbergia vernalis (Mill.) R.Gorer & J.H.Harvey, nom. rej. prop.

Distribution: Afghanistan, Armenia, Azerbaijan, Georgia, India (Kashmir), Iran (Islamic Republic
  of), Iraq, Russian Federation (the), Syrian Arab Republic (the), Tajikistan, Turkey.

Part II: Sternbergia

Sternbergia greuteriana Kamari & R.Artelari

Distribution: Greece (Crete, Kasos, Karpathos, Saria).

Sternbergia lutea (L.) Ker Gawl. ex Spreng.
Amaryllis lutea L.
Oporanthus luteus (L.) Herb.
Oporanthus luteus (L.) Herb. var. *angustifolia* Herb.
Oporanthus luteus (L.) Herb. var. *latifolia* Herb.
Sternbergia aurantiaca Dinsm.
Sternbergia lutea Ker Gawl. ex Schult. & Schult.f.

Distribution: Albania, Algeria, Azerbaijan, France, Greece (incl. Crete), Iran (Islamic Republic of),
Iraq, Israel, Italy (incl. Sardinia, Sicily), Lebanon, Spain (incl. Balearic Is.), Syrian Arab
Republic (the), Turkey, Turkmenistan, Yugoslavia (former).

Sternbergia pulchella Boiss. & Blanche

Distribution: Lebanon, Syrian Arab Republic (the).

Sternbergia schubertii Schenk

Distribution: Turkey (W).

Sternbergia sicula Tineo ex Guss.
Sternbergia lutea (L.) Spreng. subsp. *sicula* (Tineo ex Guss.) D.A.Webb
Sternbergia lutea (L.) Spreng. var. *graeca* Rchb.
Sternbergia lutea (L.) Spreng. var. *sicula* (Tineo ex Guss.) Tornab.

Distribution: Greece (incl. Aegean Is., Crete, Cyclades), Italy (incl. Sicily), Turkey (W).

PART III: COUNTRY CHECKLIST
For the genera:

Cyclamen, *Galanthus* and *Sternbergia*

TROISIÈME PARTIE: LISTE DES PAYS
Pour les genre:

Cyclamen, *Galanthus* et *Sternbergia*

PARTE III: LISTA POR PAISES
Para los géneros:

Cyclamen, *Galanthus* y *Sternbergia*

PART III: COUNTRY CHECKLIST FOR THE GENERA:
Cyclamen, Galanthus and *Sternbergia*

TROISIÈME PARTIE: LISTE PAR PAYS POUR LES GENRE:
Cyclamen, Galanthus et *Sternbergia*

PARTE III: LISTA POR PAISES PARA LOS GENEROS:
Cyclamen, Galanthus y *Sternbergia*

AFGHANISTAN / AFGHANISTAN (LE) / AFGANISTÁN (EL)

Sternbergia fischeriana (Herb.) M.Roem.

ALBANIA / ALBANIE (L') / ALBANIA

Cyclamen hederifolium Aiton
Cyclamen hederifolium Aiton var. **hederifolium** forma **albiflorum** (Jord.) Grey-Wilson
Cyclamen hederifolium Aiton var. **hederifolium** forma **hederifolium**
Galanthus nivalis L.
?Galanthus reginae-olgae Orph. subsp. **vernalis** Kamari
Sternbergia lutea (L.) Ker Gawl. ex Spreng.

ALGERIA / ALGÉRIE (L') / ALBANIA

Cyclamen africanum Boiss. & Reut.
Cyclamen persicum Mill.
Cyclamen persicum Mill. var. **persicum** forma **albidum** (Jord.) Grey-Wilson
Cyclamen persicum Mill. var. **persicum** forma **persicum**
Cyclamen repandum Sm. subsp. **repandum** var. **baborense** Debussche & Quézel
Sternbergia lutea (L.) Ker Gawl. ex Spreng.

ARMENIA / ARMÉNIE (L') / ARMENIA

Cyclamen coum Mill.
Cyclamen coum Mill. subsp. **caucasicum** (K.Koch) O.Schwarz
Galanthus alpinus Sosn.
Galanthus alpinus Sosn. var. **alpinus**
Galanthus lagodechianus Kem.-Nath.
Galanthus transcaucasicus Fomin
Sternbergia colchiciflora Waldst. & Kit.
Sternbergia fischeriana (Herb.) M.Roem.

AUSTRIA / AUTRICHE (L') / AUSTRIA

Cyclamen purpurascens Mill.
Cyclamen purpurascens Mill. forma **album** Grey-Wilson

Cyclamen purpurascens Mill. forma **purpurascens**
Galanthus nivalis L.

AZERBAIJAN / AZERBAÏDJAN (L') / AZERBAIYÁN

Cyclamen coum Mill.
Cyclamen coum Mill. subsp. **caucasicum** (K.Koch) O.Schwarz
Galanthus alpinus Sosn.
Galanthus alpinus Sosn. var. **alpinus**
Galanthus lagodechianus Kem.-Nath.
Galanthus transcaucasicus Fomin
Sternbergia colchiciflora Waldst. & Kit.
Sternbergia fischeriana (Herb.) M.Roem.
Sternbergia lutea (L.) Ker Gawl. ex Spreng.

BELGIUM / BELGIQUE (LA) / BÉLGICA

Galanthus nivalis L. (1)

BULGARIA / BULGARIE (LA) / BULGARIA

Cyclamen coum Mill.
Cyclamen coum Mill. subsp. **coum** forma **coum**
Cyclamen hederifolium Aiton
Cyclamen hederifolium Aiton var. **hederifolium** forma **albiflorum** (Jord.) Grey-Wilson
Cyclamen hederifolium Aiton var. **hederifolium** forma **hederifolium**
Galanthus elwesii Hook.f.
Galanthus gracilis Čelak.
Galanthus nivalis L.
Sternbergia colchiciflora Waldst. & Kit.

CYPRUS / CHYPRE / CHIPRE

Cyclamen cyprium Kotschy
Cyclamen graecum Link
Cyclamen graecum Link subsp. **anatolicum** Ietsw.
Cyclamen persicum Mill.
Cyclamen persicum Mill. var. **persicum** forma **albidum** (Jord.) Grey-Wilson
Cyclamen persicum Mill. var. **persicum** forma **persicum**

CZECH REPUBLIC (THE) / RÉPUBLIQUE TCHÈQUE (LA) / REPÚBLICA CHECA (LA)

Cyclamen purpurascens Mill.
Cyclamen purpurascens Mill. forma **album** Grey-Wilson
Cyclamen purpurascens Mill. forma **purpurascens**

Part III: Country Checklist / Liste par Pays / Lista por Paises

Galanthus nivalis L.

DENMARK / DANEMARK (LE) / DINAMARCA

Galanthus nivalis L. (2)

FRANCE / FRANCE (LA) / FRANCIA

Cyclamen balearicum Willk.
Cyclamen hederifolium Aiton
Cyclamen hederifolium Aiton var. hederifolium forma albiflorum (Jord.) Grey-Wilson
Cyclamen hederifolium Aiton var. hederifolium forma hederifolium
Cyclamen purpurascens Mill.
Cyclamen purpurascens Mill. forma album Grey-Wilson
Cyclamen purpurascens Mill. forma purpurascens
Cyclamen repandum Sm.
Cyclamen repandum Sm. subsp. repandum var. repandum forma album Grey-Wilson
Cyclamen repandum Sm. subsp. repandum var. repandum forma repandum
Galanthus nivalis L. (3)
Sternbergia lutea (L.) Ker Gawl. ex Spreng.

GEORGIA / GÉORGIE (LA) / GEORGIA

Cyclamen colchicum (Albov) Albov
Cyclamen coum Mill.
Cyclamen coum Mill. subsp. caucasicum (K.Koch) O.Schwarz
Galanthus alpinus Sosn.
Galanthus alpinus Sosn. var. alpinus
Galanthus krasnovii A.P.Khokhr.
Galanthus lagodechianus Kem.-Nath.
Galanthus platyphyllus Traub & Moldenke
Galanthus rizehensis Stern
Galanthus woronowii Losinsk.
Sternbergia colchiciflora Waldst. & Kit.
Sternbergia fischeriana (Herb.) M.Roem.

GERMANY / ALLEMAGNE (L') / ALEMANIA

Cyclamen purpurascens Mill.
Cyclamen purpurascens Mill. forma album Grey-Wilson
Cyclamen purpurascens Mill. forma purpurascens
Galanthus nivalis L. (3)

GREECE / GRÈCE (LA) / GRECIA

Cyclamen creticum (Dörfl.) Hildebr.

58

Cyclamen creticum (Dörfl.) Hildebr. forma **creticum**
Cyclamen creticum (Dörfl.) Hildebr. forma **pallide-rosea** Grey-Wilson
Cyclamen graecum Link
Cyclamen graecum Link subsp. **anatolicum** Ietsw.
Cyclamen graecum Link subsp. **graecum** forma **album** R.Frank
Cyclamen graecum Link subsp. **graecum** forma **graecum**
Cyclamen graecum Link subsp. **mindleri** (Heldr.) A.P.Davis & Govaerts
Cyclamen hederifolium Aiton
Cyclamen hederifolium Aiton var. **confusum**
Cyclamen hederifolium Aiton var. **hederifolium** forma **albiflorum** (Jord.) Grey-Wilson
Cyclamen hederifolium Aiton var. **hederifolium** forma **hederifolium**
Cyclamen persicum Mill.
Cyclamen persicum Mill. var. **persicum** forma **albidum** (Jord.) Grey-Wilson
Cyclamen persicum Mill. var. **persicum** forma **persicum**
Cyclamen repandum Sm.
Cyclamen repandum Sm. subsp. **peloponnesiacum** var. **peloponnesiacum** (Grey-Wilson) Grey-Wilson
Cyclamen repandum Sm. subsp. **peloponnesiacum** var. **vividum** (Grey-Wilson) Grey-Wilson
Cyclamen repandum Sm. subsp. **repandum** var. **repandum** forma **album** Grey-Wilson
Cyclamen repandum Sm. subsp. **rhodense** (Meikle) Grey-Wilson
Galanthus elwesii Hook.f.
Galanthus gracilis Čelak.
Galanthus ikariae Baker
Galanthus nivalis L.
Galanthus peshmenii A.P.Davis & C.D.Brickell
Galanthus reginae-olgae Orph.
Galanthus reginae-olgae Orph. subsp. **reginae-olgae**
Galanthus reginae-olgae Orph. subsp. **vernalis** Kamari
Sternbergia clusiana (Ker Gawl.) Ker Gawl. ex Spreng.
Sternbergia colchiciflora Waldst. & Kit.
Sternbergia greuteriana Kamari & R.Artelari
Sternbergia lutea (L.) Ker Gawl. ex Spreng.
Sternbergia sicula Tineo ex Guss.

HUNGARY / HONGRIE (LA) / HUNGRÍA

Cyclamen purpurascens Mill.
Cyclamen purpurascens Mill. forma **album** Grey-Wilson
Cyclamen purpurascens Mill. forma **purpurascens**
Galanthus nivalis L.
Sternbergia colchiciflora Waldst. & Kit.

INDIA (KASHMIR) / INDE (L') (KASHMIR) / INDIA (LA) (KASHMIR)

Sternbergia fischeriana (Herb.) M.Roem.

IRAN (ISLAMIC REPUBLIC OF) / IRAN (RÉPUBLIQUE ISLAMIQUE D') / IRÁN (REPÚBLICA ISLÁMICA DEL)

Cyclamen coum Mill.
Cyclamen coum Mill. subsp. **elegans** (Boiss. & Buhse) Grey-Wilson
? Galanthus alpinus Sosn.
? Galanthus alpinus Sosn. var. **alpinus**
Galanthus transcaucasicus Fomin
Sternbergia clusiana (Ker Gawl.) Spreng.
Sternbergia colchiciflora Waldst. & Kit.
Sternbergia fischeriana (Herb.) M.Roem.
Sternbergia lutea (L.) Ker Gawl. ex Spreng.

IRAQ / IRAQ (L') / IRAQ (EL)

Sternbergia clusiana (Ker Gawl.) Spreng.
Sternbergia fischeriana (Herb.) M.Roem.
Sternbergia lutea (L.) Ker Gawl. ex Spreng.

ISRAEL / ISRAËL / ISRAEL

Cyclamen coum Mill.
Cyclamen coum Mill. subsp. **coum** forma **coum**
Cyclamen persicum Mill.
Cyclamen persicum Mill. var. **autumnale** Grey-Wilson
Cyclamen persicum Mill. var. **persicum** forma **albidum** (Jord.) Grey-Wilson
Cyclamen persicum Mill. var. **persicum** forma **persicum**
? Galanthus fosteri Baker
Sternbergia clusiana (Ker Gawl.) Spreng.
Sternbergia colchiciflora Waldst. & Kit.
Sternbergia lutea (L.) Ker Gawl. ex Spreng.

ITALY / ITALIE (L') / ITALIA

Cyclamen hederifolium Aiton
Cyclamen hederifolium Aiton var. **confusum**
Cyclamen hederifolium Aiton var. **hederifolium** forma **albiflorum** (Jord.) Grey-Wilson
Cyclamen hederifolium Aiton var. **hederifolium** forma **hederifolium**
Cyclamen purpurascens Mill.
Cyclamen purpurascens Mill. forma **album** Grey-Wilson
Cyclamen purpurascens Mill. forma **purpurascens**
Cyclamen repandum Sm.
Cyclamen repandum Sm. subsp. **repandum** var. **repandum** forma **album** Grey-Wilson
Cyclamen repandum Sm. subsp. **repandum** var. **repandum** forma **repandum**
Galanthus nivalis L.
Galanthus reginae-olgae Orph.
Galanthus reginae-olgae Orph. subsp. **reginae-olgae**
Galanthus reginae-olgae Orph. subsp. **vernalis** Kamari

Sternbergia colchiciflora Waldst. & Kit.
Sternbergia lutea (L.) Ker Gawl. ex Spreng.
Sternbergia sicula Tineo ex Guss.

JORDAN / JORDANIE (LA) / JORDANIA

Cyclamen persicum Mill.
Cyclamen persicum Mill. var. **persicum** forma **albidum** (Jord.) Grey-Wilson
Cyclamen persicum Mill. var. **persicum** forma **persicum**
? Cyclamen persicum Mill. var. **persicum** forma **puniceum** Grey-Wilson
? Cyclamen persicum Mill. var. **persicum** forma **roseum** Grey-Wilson, nom. provis.
Galanthus fosteri Baker
Sternbergia clusiana (Ker Gawl.) Spreng.

LEBANON / LIBAN (LE) / LÍBANO (EL)

Cyclamen coum Mill.
Cyclamen coum Mill. subsp. **coum** forma **coum**
Cyclamen libanoticum Hildebr.
Cyclamen persicum Mill.
Cyclamen persicum Mill. var. **persicum** forma **albidum** (Jord.) Grey-Wilson
Cyclamen persicum Mill. var. **persicum** forma **persicum**
? Cyclamen persicum Mill. var. **persicum** forma **puniceum** Grey-Wilson
? Cyclamen persicum Mill. var. **persicum** forma **roseum** Grey-Wilson nom. provis.
Galanthus fosteri Baker
Sternbergia clusiana (Ker Gawl.) Spreng.
Sternbergia lutea (L.) Ker Gawl. ex Spreng.
Sternbergia pulchella Boiss. & Blanche

LIBYAN ARAB JAMAHIRIYA (THE) / JAMAHIRIYA ARABE LIBYENNE (LA) / JAMAHIRIYA ARABE LIBIA (LA)

Cyclamen rohlfsianum Asch.

THE NETHERLANDS (THE) / PAYS-BAS (LES) PAÍSES BAJOS (LOS)

Galanthus elwesii Hook.f. (2)
Galanthus nivalis L. (2)
Galanthus woronowii Losinsk. (2)

POLAND / POLOGNE (LA) / POLONIA

Cyclamen purpurascens Mill.
Cyclamen purpurascens Mill. forma **album** Grey-Wilson
Cyclamen purpurascens Mill. forma **purpurascens**
Galanthus nivalis L.

REPUBLIC OF MOLDOVA (THE) / RÉPUBLIQUE DE MOLDOVA (LA) / REPÚBLIC DE MOLDOVA (LA)

Galanthus nivalis L.

ROMANIA / ROUMANIE (LA) / RUMANIA

Cyclamen purpurascens Mill. (2)
Cyclamen purpurascens Mill. forma purpurascens (2)
Galanthus gracilis Čelak. (2)
Galanthus nivalis L.
Galanthus plicatus M.Bieb.
Galanthus plicatus M.Bieb. subsp. plicatus
Sternbergia colchiciflora Waldst. & Kit.

RUSSIAN FEDERATION (THE) / FÉDÉRATION DE RUSSIE (LA) / FEDERACIÓN DE RUSIA (LA)

Cyclamen coum Mill.
Cyclamen coum Mill. subsp. caucasicum (K.Koch) O.Schwarz
Cyclamen coum Mill. subsp. coum forma coum
Cyclamen purpurascens Mill. (2)
Cyclamen purpurascens Mill. forma purpurascens (2)
Galanthus alpinus Sosn.
Galanthus alpinus Sosn. var. alpinus
Galanthus alpinus var. bortkewitschianus (Koss) A.P.Davis
Galanthus angustifolius Koss
Galanthus lagodechianus Kem.-Nath.
Galanthus platyphyllus Traub & Moldenke
Galanthus plicatus M.Bieb.
Galanthus plicatus M.Bieb. subsp. plicatus
Galanthus rizehensis Stern
Galanthus woronowii Losinsk.
Sternbergia colchiciflora Waldst. & Kit.
Sternbergia fischeriana (Herb.) M.Roem.

SLOVAKIA / SLOVAQUIE (LA) / ESLOVAQUIA

Cyclamen purpurascens Mill.
Cyclamen purpurascens Mill. forma album Grey-Wilson
Cyclamen purpurascens Mill. forma purpurascens
Galanthus nivalis L.

SOMALIA / SOMALIE (LA) / SOMALIA

Cyclamen somalense Thulin & Warfa

SPAIN / ESPAGNE (L') / ESPAÑA

Cyclamen balearicum Willk.
? Galanthus nivalis L.
Sternbergia colchiciflora Waldst. & Kit.
Sternbergia lutea (L.) Ker Gawl. ex Spreng.

SWEDEN / SUÈDE (LA) / SUCCIA

Galanthus nivalis L. (2)

SWITZERLAND / SUISSE (LA) / SUIZA

Cyclamen hederifolium Aiton
Cyclamen hederifolium Aiton var. **hederifolium** forma **albiflorum** (Jord.) Grey-Wilson
Cyclamen hederifolium Aiton var. **hederifolium** forma **hederifolium**
Cyclamen purpurascens Mill.
Cyclamen purpurascens Mill. forma **album** Grey-Wilson
Cyclamen purpurascens Mill. forma **purpurascens**
Cyclamen repandum Sm.
? Cyclamen repandum Sm. subsp. **repandum** var. **repandum** forma **album** Grey-Wilson
Cyclamen repandum Sm. subsp. **repandum** var. **repandum** forma **repandum**
Galanthus nivalis L.

SYRIAN ARAB REPUBLIC (THE) / RÉPUBLIQUE ARABE SYRIENNE (LA) / REPÚBLICA ARABE SIRIA (LA)

Cyclamen coum Mill.
Cyclamen coum Mill. subsp. **coum** forma **coum**
Cyclamen persicum Mill.
Cyclamen persicum Mill. var. **persicum** forma **albidum** (Jord.) Grey-Wilson
Cyclamen persicum Mill. var. **persicum** forma **persicum**
Cyclamen persicum Mill. var. **persicum** forma **puniceum** Grey-Wilson
Cyclamen persicum Mill. var. **persicum** forma **roseum** Grey-Wilson, nom. provis.
Galanthus fosteri Baker
Sternbergia clusiana (Ker Gawl.) Spreng.
Sternbergia fischeriana (Herb.) M.Roem.
Sternbergia lutea (L.) Ker Gawl. ex Spreng.
Sternbergia pulchella Boiss. & Blanche

TAJIKISTAN / TADJIKISTAN (LE) / TAYIKISTÁN

Sternbergia fischeriana (Herb.) M.Roem.

Part III: Country Checklist / Liste par Pays / Lista por Paises

TUNISIA / TUNISIE (LA) / TÚNEZ

Cyclamen africanum Boiss. & Reut.
Cyclamen persicum Mill.
Cyclamen persicum Mill. var. **persicum** forma **albidum** (Jord.) Grey-Wilson
Cyclamen persicum Mill. var. **persicum** forma **persicum**

TURKEY / TURQUIE (LA) / TURQUÍA

Cyclamen cilicium Boiss. & Heldr.
Cyclamen cilicium Boiss. & Heldr. forma **album** E.Frank & Koenen
Cyclamen cilicium Boiss. & Heldr. forma **cilicium**
Cyclamen coum Mill.
Cyclamen coum Mill. subsp. **caucasicum** (K.Koch) O.Schwarz
Cyclamen coum Mill. subsp. **coum** forma **albissimum** R.H.Bailey, Koenen, Lillywh. &
P.J.M.Moore
Cyclamen coum Mill. subsp. **coum** forma **coum**
Cyclamen coum Mill. subsp. **coum** forma **pallidum** Grey-Wilson
Cyclamen graecum Link
Cyclamen graecum Link subsp. **anatolicum** Ietsw.
Cyclamen hederifolium Aiton
Cyclamen hederifolium Aiton var. **hederifolium** forma **albiflorum** (Jord.) Grey-Wilson
Cyclamen hederifolium Aiton var. **hederifolium** forma **hederifolium**
Cyclamen intaminatum (Meikle) Grey-Wilson
Cyclamen mirabile Hildebr.
Cyclamen mirabile Hildebr. forma **mirabile**
Cyclamen mirabile Hildebr. forma **niveum** Grey-Wilson
Cyclamen parviflorum Poped.
Cyclamen parviflorum Poped. var. **parviflorum**
Cyclamen parviflorum Poped. var. **subalpinum** Grey-Wilson
Cyclamen persicum Mill.
Cyclamen persicum Mill. var. **persicum** forma **albidum** (Jord.) Grey-Wilson
Cyclamen persicum Mill. var. **persicum** forma **persicum**
Cyclamen pseudibericum Hildebr.
Cyclamen pseudibericum Hildebr. forma **pseudibericum**
Cyclamen pseudibericum Hildebr. forma **roseum** Grey-Wilson
Cyclamen trochopteranthum O.Schwarz
Cyclamen trochopteranthum O.Schwarz forma **leucanthum** Grey-Wilson
Cyclamen trochopteranthum O.Schwarz forma **trochopteranthum**
Galanthus alpinus Sosn.
Galanthus alpinus Sosn. var. **alpinus**
Galanthus cilicicus Baker
Galanthus elwesii Hook.f.
Galanthus fosteri Baker
Galanthus gracilis Čelak.
Galanthus koenenianus Lobin, C.D.Brickell & A.P.Davis
Galanthus krasnovii A.P.Khokhr.
Galanthus nivalis L.
Galanthus peshmenii A.P.Davis & C.D.Brickell
Galanthus plicatus M.Bieb.
Galanthus plicatus M.Bieb. subsp. **byzantinus** (Baker) D.A.Webb

Galanthus plicatus M.Bieb. subsp. **plicatus**
Galanthus rizehensis Stern
Galanthus woronowii Losinsk.
Sternbergia candida B.Mathew & T.Baytop
Sternbergia clusiana (Ker Gawl.) Spreng.
Sternbergia colchiciflora Waldst. & Kit.
Sternbergia fischeriana (Herb.) M.Roem.
Sternbergia lutea (L.) Ker Gawl. ex Spreng.
Sternbergia schubertii Schenk
Sternbergia sicula Tineo ex Guss.

TURKMENISTAN / TURKMÉNISTAN (LE) / TURKMENISTÁN

Sternbergia lutea (L.) Ker Gawl. ex Spreng.

UKRAINE / UKRAINE (L') / UCRANIA

Galanthus elwesii Hook.f.
Galanthus gracilis Čelak.
Galanthus nivalis L.

UNITED KINGDOM OF GREAT BRITAIN AND NORTHERN IRELAND / ROYAUME-UNI DE GRANDE-BRETAGNE ET D'IRLANDE DU NORD (LE) / REINO UNIDO DE GRAN BRETANA E IRLANDA DEL NORTE (EL)

Cyclamen hederifolium Aiton (2)
Cyclamen hederifolium Aiton var. **hederifolium** forma **hederifolium** (2)
Galanthus alpinus (2)
Galanthus elwesii Hook.f. (2)
Galanthus nivalis L. (2)
Galanthus plicatus M.Bieb. (2)
Galanthus plicatus M.Bieb. subsp. **plicatus** (2)
Galanthus woronowii Losinsk. (2)

YUGOSLAVIA (FORMER) / YOUGOSLAVIE (LA) / YUGOSLAVIA

Cyclamen hederifolium Aiton
Cyclamen hederifolium Aiton var. **hederifolium** forma **albiflorum** (Jord.) Grey-Wilson
Cyclamen hederifolium Aiton var. **hederifolium** forma **hederifolium**
Cyclamen purpurascens Mill.
Cyclamen purpurascens Mill. forma **album** Grey-Wilson
Cyclamen purpurascens Mill. forma **purpurascens**
Cyclamen repandum Sm.
Cyclamen repandum Sm. subsp. **repandum** var. **repandum** forma **album** Grey-Wilson
Cyclamen repandum Sm. subsp. **repandum** var. **repandum** forma **repandum**
Galanthus elwesii Hook.f.
Galanthus nivalis L.
Galanthus reginae-olgae Orph.

Galanthus reginae-olgae Orph. subsp. **vernalis** Kamari
Sternbergia colchiciflora Waldst. & Kit.
Sternbergia lutea (L.) Ker Gawl. ex Spreng.

OF HORTICULTURAL/GARDEN ORIGIN: NOT FOUND IN THE WILD

D'ORIGINE HORTICOLE/DE JARDIN: N'EXISTE PAS DANS LA NATURE

DE ORIGEN HORTÍCOLA/JARDÍN: NO SE ENCUENTRA EN LA NATURALEZA

Cyclamen × **atkinsii** T.Moore (? *Cyclamen coum* × *Cyclamen persicum*)
Cyclamen × **drydeniae** Grey-Wilson (*Cyclamen coum* × *Cyclamen trochopteranthum*)
Cyclamen × **hildebrandii** O.Schwarz (*Cyclamen africanum* × *Cyclamen hederifolium*)
Cyclamen × **meiklei** Grey-Wilson (*Cyclamen creticum* × *Cyclamen repandum*)
Cyclamen × **saundersiae** Grey-Wilson (*Cyclamen balearicum* × *Cyclamen repandum*)
Cyclamen × **schwarzii** Grey-Wilson (*Cyclamen libanoticum* × *Cyclamen pseudibericum*)
Cyclamen × **wellensiekii** Ietsw. (*Cyclamen cyprium* × *Cyclamen libanoticum*)
Cyclamen × **whiteae** Grey-Wilson (*Cyclamen graecum* × *Cyclamen hederifolium*)
Galanthus × **allenii** Baker (?*Galanthus alpinus* × *Galanthus woronowii* Losinsk.)
Galanthus × **grandiflorus** Baker (? *Galanthus nivalis* × *Galanthus plicatus*)
Galanthus nivalis L. **'Flore Pleno'**

Key: 1=probably introduced; 2=introduced and naturalized; 3=introduced in certain areas of this country, but otherwise native.

Clé: 1=probablement introduit; 2=introduit et naturalisé; 3=introduit dans certaines régions de ce pays; autrement, natif.

Clave: 1=probablemente introducida; 2=introducida y naturalizada; 3=introducida en ciertas zonas de este país; de otro modo se trata de una especie nativa.

Annex I: All names, with author and place of publication
Annex I: Tous les noms avec les auteurs et le lieu de publication
Anexo I: Todos los nombres, con el autor y el lugar de publicacion

ANNEX I: LIST OF ALL NAMES, WITH AUTHOR AND PLACE OF PUBLICATION
Ordered alphabetically on all names:

Cyclamen, Galanthus and *Sternbergia*

ANNEXE I: TOUS LES NOMS AVEC LES AUTEURS ET LE LIEU DE PUBLICATION
Par ordre alphabétique de tous les noms:

Cyclamen, Galanthus et *Sternbergia*

ANEXO I: TODOS LOS NOMBRES, CON EL AUTOR Y EL LUGAR DE PUBLICACION
Presentados por orden alfabético: todos los nombres

Cyclamen, Galanthus y *Sternbergia*

Annex I: All names with author and place of publication - *Cyclamen*
Annexe I: Tous les noms avec les auteurs et le lieu de publication - *Cyclamen*
Anexo I: Todos los nombres, con el autor y el lugar de publicacion - *Cyclamen*

ALPHABETICAL LISTING OF ALL NAMES, WITH AUTHOR AND PLACE OF PUBLICATION
Accepted names in **bold**

CYCLAMEN

Cyclamen abchasicum (Medw. ex Kusn.) Kolak., Fl. Abkhazya 3: 274 (1948)
Cyclamen adzharicum Poped. in Bot. Mater. Gerb. Glavn. Bot. Sada SSR 13: 189 (1950).
Cyclamen aedirhizum Jord. in Jord. & Fourr., Ic. Fl. Eur. 3: 21 (1903).
Cyclamen aegineticum Hildebr. in Gartenflora 57: 296 (1908).
Cyclamen aestivum Rchb., Fl. Germ. Excurs. 1: 407 (1830).
Cyclamen africanum Boiss. & Reut., Pugill. Pl. Afr. Bot. Hispañ.: 75 (1852).
Cyclamen albidum Jord. in Jord. & Fourr., Ic. Fl. Eur. 3: 23 (1903).
Cyclamen albiflorum Jord. in Jord. & Fourr., Ic. Fl. Eur. 3: 20 (1903).
Cyclamen aleppicum Fisch. ex Hoffmanns., Verz. Pfl.-Kult.: 54 (1824).
Cyclamen aleppicum Hoffmanns. subsp. *puniceum* Glasau in Planta 30: 545 (1939).
Cyclamen algeriense Jord. in Jord. & Fourr., Ic. Fl. Eur. 3: 22 (1903).
Cyclamen alpinum hort. Dammann ex Sprenger in Gartenflora 41: 526 (1892).
Cyclamen alpinum sens. Turrill in Curtis's Bot. Mag. 174: t. 437 (1963).
Cyclamen alpinum hort. Dammann ex Sprenger var. *album* hort. Dammann ex Sprenger in Gartenflora 41: 526 (1892).
Cyclamen ambiguum O.Schwarz in Feddes Repert. Spec. Nov. Regni Veg. 58: 275 (1955).
Cyclamen angulare Jord. in Jord. & Fourr., Ic. Fl. Eur. 3: 19 (1903).
Cyclamen antilochium Decne in Rev. Hortic. ser. 4, 5: 23 (1855).
Cyclamen apiculatum Jord. in Jord. & Fourr., Ic. Fl. Eur. 3: 15, t. 402 (1903).
Cyclamen atkinsii Glasau in Planta 30: 516 (1939), non T.Moore
Cyclamen atkinsii hort., non T.Moore. No indication of earliest usage.
Cyclamen × atkinsii T.Moore in A. Henfrey et al., Gard. Compan. Florists' Guide 1: 89 (1852).
Cyclamen autumnale Boos, Schönbrunn's Fl.: 45 (1816).
Cyclamen baborense Batt. in Marie, Herbier de l'Afrique du Nord [in sched.], without date.
Cyclamen balearicum Willk. in Oesterr. Bot. Z. 25: 111 (1875).
Cyclamen breviflorum Jord. in Jord. & Fourr., Ic. Fl. Eur. 3: 18 (1903).
Cyclamen brevifrons Jord. in Jord. & Fourr., Ic. Fl. Eur. 3: 15, t. 401 (1903).
Cyclamen calcareum Kolak. in Izv. Glavn. Bot. Sada SSSR 3: 83 (1949).
Cyclamen caucasicum Willd. ex Boiss., Fl. Orient. 4: 11 (1875).
Cyclamen caucasicum Willd. ex Steven, Bull. in Soc. Imp. Naturaliste Moscou 30: 327 (1857).
Cyclamen cilicium Boiss. & Heldr. [forma **cilicium**] in Boiss., Diagn. ser. 1, 11: 78 (1843).
Cyclamen cilicium Boiss. & Heldr. forma **album** E.Frank & Koenen in R.Frank & E.Frank in Bull. Alp. Gard. Soc. Gr. Brit. 51: 150 (1983).
Cyclamen cilicium Boiss. & Heldr. var. *alpinum* hort. [?unpublished].
Cyclamen cilicium Boiss. & Heldr. var. *intaminatum* Meikle in Notes Roy. Bot. Gard. Edinburgh 36: 2 (1978); Meikle in Davis, Fl. Turkey 6: 130 (1978).
Cyclamen cilicium Boiss. & Heldr. var. [*sic*] Turrill in Curtis's Bot. Mag. 171: t. 307 (1957).
Cyclamen cilicium Hildebr., Gattung *Cyclamen*: 36 (1898).
Cyclamen circassicum Poped. in Bot. Zhurn. (Moscow & Leningrad) 33: 226 (1948).
Cyclamen clusii Lindl., Bot. Reg. 12: t. 1013 (1826).
Cyclamen colchicum (Albov) Albov in Wien Ill. Gart.-Zeitung 23: 7 (1898).
Cyclamen commutatum O.Schwarz & Lepper in Feddes Repert. Spec. Nov. Regni Veg. 69: 91 (1964).
Cyclamen cordifolium Stokes, Bot. Mat. Med. 1: 295 (1812).
Cyclamen coum Mill. [subsp. **coum** forma **coum**], Gard. Dict. edn. 8, n. 6 (1768).
Cyclamen coum sens. Rchb., Fl. Germ. Excurs. 1: 406 (1830), non Mill.

Annex I: All names with author and place of publication - *Cyclamen*
Annexe I: Tous les noms avec les auteurs et le lieu de publication - *Cyclamen*
Anexo I: Todos los nombres, con el autor y el lugar de publicacion - *Cyclamen*

Cyclamen coum Mill. subsp. *alpinum* (hort. Dammann ex Sprenger) O.Schwarz in Feddes Repert. Spec. Nov. Regni Veg. 58: 250 (1955).

Cyclamen coum Mill. subsp. **caucasicum** (K.Koch) O.Schwarz in Feddes Repert. Spec. Nov. Regni Veg. 58: 250 (1955).

Cyclamen coum Mill. subsp. **coum** forma **albissimum** R.H.Bailey, Koenen, Lillywh. & P.J.M.Moore in Cyclamen Soc. J. 13(1): 27 (1989).

Cyclamen coum Mill. subsp. **coum** forma **pallidum** Grey-Wilson, *Cyclamen*: 174 (1997).

Cyclamen coum Mill. subsp. **elegans** (Boiss. & Buhse) Grey-Wilson *Cyclamen*: 174 (1997).

Cyclamen coum Mill. subsp. *hiemale* (Hildebr.) O.Schwarz in Feddes Repert. Spec. Nov. Regni Veg. 58: 249 (1955).

Cyclamen coum Mill. var. *abchasicum* Medw. ex Kusn. in Mat. Fl. Cauc. 4(1): 167 (1902) [there is some uncertainty as to whether this should be Medw. in or ex Kusn., as Schwarz (1955) gives it as Medw. in Kusn, and Popedimova (1952) takes up Medw. instead of Kusn. - reference not seen].
[il y a une certaine incertitude de savoir si ceci devrait être Medw. dans ou Kusn. ex, comme Schwarz (1955) le donne comme Medw. dans Kusn, et Popedimova (1952) prend Medw. au lieu de Kusn. - mettent en référence non vu].
[hay una cierta incertidumbre si éste debe ser Medw. en o Kusn. ex, como Schwarz (1955) lo da como Medw. en Kusn, y Popedimova (1952) toma Medw. en vez de Kusn. - se refieren no visto].

Cyclamen coum Mill. var. *caucasicum* (K.Koch) Meikle in Davis, Fl. Turkey 6: 133 (1978).

Cyclamen coum Mill. var. *ibericum* Boiss., Fl. Orient. 4: 11 (1879).

Cyclamen crassifolium Hildebr. in Beih. Bot. Centralbl. 22(2): 195, t. 6 (1907).

Cyclamen creticum (Dörfl.) Hildebr. [forma **creticum**]in Beih. Bot. Centralbl. 19(2): 367 (1906).

Cyclamen creticum (Dörfl.) Hildebr. forma **pallide-roseum** Grey-Wilson, *Cyclamen*: 174 (1997).

Cyclamen cyclaminus Bedevian, Illustr. Polyglot. Dict.: 218 (1936).

Cyclamen cyclophyllum Jord. in Jord. & Fourr., Ic. Fl. Eur. 3: 18 (1903).

Cyclamen cyprium Glasau in Planta 30: 537 (1939), non Kotschy.

Cyclamen cyprium Kotschy in Unger & Kotschy, Ins. Cypern: 295 (1865).

Cyclamen cyprium Sibth. in Walpole, Travels: 25 (1820).

Cyclamen cypro-graecum E.Mutch & N.Mutch in Bull. Alpine Gard. Soc. Gr. Brit. 23: 164 (1954).

Cyclamen deltoideum Tausch in Flora 12(2): 667 (1829).

Cyclamen × **drydeniae** Grey-Wilson, *Cyclamen*: 174 (1997), as "**drydenii**".

Cyclamen durostoricum Pantu & Solacolu in Bull. Sect. Sci. Acad. Roumaine. 9: 23 (1924).

Cyclamen elegans Boiss. & Buhse in Nouv. Mém. Soc. Imp. Naturalistes Moscou 12: 145 (1860).

Cyclamen eucardium Jord. in Jord. & Fourr., Ic. Fl. Eur. 3: 16 (1903).

Cyclamen europaeum L., Sp. Pl. 1: 145 (1753) - (includes more than one taxon - *pro parte*).

Cyclamen europaeum L. subsp. *orbiculatum* O.Schwarz in Gartenflora n.s., 1: 16 (1938), non *C. orbiculatum* Mill.

Cyclamen europaeum L. subsp. *orbiculatum* (Mill.) O.Schwarz var. *immaculatum* Hrabetova in Ceskoslov. Bot. Listy 3: 35 (1950).

Cyclamen europaeum L. subsp. *ponticum* (Albov) O.Schwarz in Gartenflora n.s., 1: 16 (1938).

Cyclamen europaeum L. var. *caucasicum* K.Koch in Linnaea 23: 619 (1849).

Cyclamen europaeum L. var. *colchicum* Albov in Trudy Tbilissk. Bot. Sada. 1: 166 (1895).

Cyclamen europaeum L. var. *ponticum* Albov in Bull. Herb. Boiss. 2: 254 (1894).

Cyclamen europaeum L. var. *typicum* Albov in Bull. Herb. Boiss. 2: 254 (1894).

Cyclamen europaeum Pall. in Nova Acta Acad. Sci. Imp. Petrop. Hist. Acad. 10: 306 (1796).

Cyclamen europaeum Savi, Fl. Pis. 1: 213 (1798).

Cyclamen europaeum sens. Aiton, Hort. Kew. edn. 1, 196 (1789).

Cyclamen europaeum sens. Albov in Bull. Herb. Boiss. 2: 254 (1854).

Cyclamen europaeum sens. Mill., Gard. Dict. edn. 8, no. 1. (1768).

Cyclamen europaeum Sm. in Engl., Bot. t. 548 [reference not found, as listed in *Index Kewensis*]

Cyclamen fatrense Halda & Soják in Cas. Nár. Muz. Odd. Prír. 140(1-2): 64 (1971).

Cyclamen ficariifolium Rchb., Fl. Germ. Excurs. 1: 407 (1830).

Annex I: All names with author and place of publication - *Cyclamen*
Annexe I: Tous les noms avec les auteurs et le lieu de publication - *Cyclamen*
Anexo I: Todos los nombres, con el autor y el lugar de publicacion - *Cyclamen*

Cyclamen floridum Salisb., Prodr. Stirp. Chap. Allerton: 119 (1796).
Cyclamen gaidurowryssii Glasau in Planta 31: 539 (1939).
Cyclamen gaydurowryssii (orththographical variant) Glasau in Planta 31: 539 (1939).
Cyclamen graecum Link [subsp. **graecum** forma **graecum**] in Linnaea 9: 573 (1834).
Cyclamen graecum Link subsp. **anatolicum** Ietsw. in Cyclamen Soc. J. 14(2): 51 (1990).
Cyclamen graecum Link subsp. *candicum* Ietsw. in Cyclamen Soc. J. 14(2): 50 (1990).
Cyclamen graecum Link subsp. **graecum** forma **album** R.Frank & E.Frank in Bull. Alp. Gard. Soc. Gr. Brit. 50(3): 251 (photo. on p. 250) (1982).
Cyclamen graecum Link subsp. **mindleri** (Heldr.) A.P.Davis & Govaerts, see page 129.
Cyclamen hastatum Tausch in Flora 12: 668 (1829).
Cyclamen hederaceum Sieber ex Steud., Nomencl. Bot. edn. 2, 1: 458 (1821).
Cyclamen hederifolium Aiton [var. **hederifolium** forma **hederifolium**], Hort. Kew. edn. 1, 196 (1789).
Cyclamen hederifolium Kotschy in Unger & Kotschy, Ins. Cypern: 295 (1865).
Cyclamen hederifolium Aiton var. **confusum** Grey-Wilson, *Cyclamen*: 174 (1997).
Cyclamen hederifolium Aiton subsp. *balearicum* O.Schwarz in Gartenflora n.s., 1: 22 (1938).
Cyclamen hederifolium Aiton subsp. *creticum* (Dörfl.) O.Schwarz in Gartenflora n.s., 1: 22 (1938).
Cyclamen hederifolium Aiton subsp. *romanum* (Griseb.) O.Schwarz in Gartenflora n.s., 1: 22 (1938).
Cyclamen hederifolium Aiton var. **hederifolium** forma **albiflorum** (Jord.) Grey-Wilson, *Cyclamen*: 174 (1997).
Cyclamen hederifolium Sibth. & Sm., Fl. Graec. Prod. 1: 128 (1813).
Cyclamen hederifolium Sims in Bot. Mag. 25: t. 1001 (1807).
Cyclamen hederifolium Willd., Sp. Pl. 1(2): 810 (1798).
Cyclamen hiemale Hildebr. in Gartenflora 53: 70 (1904).
Cyclamen × **hildebrandii** O.Schwarz in Feddes Repert. Spec. Nov. Regni Veg. 58: 280 (1955).
Cyclamen holochlorum Jord. in Jord. & Fourr., Ic. Fl. Eur. 3: 19 (1903).
Cyclamen hyemale Salisb., Prodr. Stirp. Chap. Allerton: 118 (1796).
Cyclamen ibericum Goldie ex G.Don in Sweet, Hort. Brit. edn. 3: 560 (1839).
Cyclamen ibericum Lem., Jard. Fleur. 3: t. 297 (1853), excl. descr., non Goldie.
Cyclamen ibericum Steven ex Boiss., Fl. Orient. 4: 11 (1875).
Cyclamen ibericum T.Moore in A. Henfrey et al., Gard. Compan. Florists' Guide 1: 90, f. 2 (1852).
Cyclamen ilicetorum Jord. in Jord. & Fourr., Ic. Fl. Eur. 3: 20 (1903).
Cyclamen immaculatum Pieri in Ionios Anthol. 5: 192 (c.1835).
Cyclamen indicum L., Sp. Pl.: 145 (1753).
Cyclamen insulare Jord. in Jord. & Fourr., Ic. Fl. Eur. 3: 20 (1903).
Cyclamen intaminatum (Meikle) Grey-Wilson, Gen. *Cyclamen*: 71 (1988).
Cyclamen intermedium Wender., Ind. Sem. Hort. Marb.: [page number not known] (1825).
Cyclamen jovis Hildebr. in Gartenflora 57: 294 (1908).
Cyclamen kusnetzovii Kotov & Czernova, Fl. RSS Ucr. 8: 521 (1958).
Cyclamen latifolium Sm. in Sibth. & Sm., Fl. Graec. 2: 71, t. 185 (1813).
Cyclamen libanoticum Hildebr. in Engl. Bot. Jahrb. 25: 477 (1898).
Cyclamen libanoticum Hildebr. subsp. *pseudibericum* (Hildebr.) Glasau in Planta 30: 523 (1939).
Cyclamen lilacinum Jord. in Jord. & Fourr., Ic. Fl. Eur. 3: 19 (1903).
Cyclamen linaerifolium DC., Fl. Fr. 3: 433 (1805).
Cyclamen littorale Sadler ex Rchb., Fl. Germ. Excurs. 1: 406 (1830).
Cyclamen lobospilum Jord. in Jord. & Fourr., Ic. Fl. Eur. 3: 17 (1903).
Cyclamen macrophyllum Sieber in Isis: 259 (1823).
Cyclamen macropus Zucc., Del. Sem. Hort. Monac.: 4 (1846).
Cyclamen maritimum Hildebr. in Gartenflora 57: 293 (1908).
Cyclamen × *marxii* Halda in Skalničky 1973(1): 28 (1973).
Cyclamen × **meiklei** Grey-Wilson, *Cyclamen*: 174 (1997).
Cyclamen miliarakesii Heldr. ex Halácsy, Consp. Fl. Graec. 3: 9 (1904).
Cyclamen miliarakesii Heldr. ex Hildebr. in Gartenflora 55: 634 (1906).

Annex I: All names with author and place of publication - *Cyclamen*
Annexe I: Tous les noms avec les auteurs et le lieu de publication - *Cyclamen*
Anexo I: Todos los nombres, con el autor y el lugar de publicacion - *Cyclamen*

Cyclamen mirabile Hildebr. [forma **mirabile**] in Beih. Bot. Centralbl. 19(2): 370 (1906).

Cyclamen mirabile Hildebr. forma **niveum** J.White & Grey-Wilson, *Cyclamen*: 174 (1997).

Cyclamen neapolitanum sens. Boiss., Fl. Orient. 4: 13 (1875).

Cyclamen neapolitanum sens. Duby in DC., Prodr. 8: 57 (1844).

Cyclamen neapolitanum Ten., Prodr. Fl. Napol. suppl. 2: 66 (1813).

Cyclamen numidicum Glasau in Planta 30: 528 (1939).

Cyclamen officinale Wender. ex Steud., Nomencl. Bot. edn. 2, 1: 458 (1841).

Cyclamen orbiculatum Mill., Gard. Dict. edn. 8, no. 5 (1768).

Cyclamen orbiculatum Mill. var. *alpinum* Saunders in Bull. Alpine Gard. Soc. Gr. Brit. 27: 49 (1959).

Cyclamen orbiculatum Mill. var. *coum* (Mill.) Door. in Meded., Landbouwhogeschool 50: 25 (1950).

Cyclamen pachylobum Jord. in Jord. & Fourr., Ic. Fl. Eur. 3: 22 (1903).

Cyclamen parviflorum Poped. [var. **parviflorum**] in Bot. Mater. Gerb. Bot. Inst. Komarova Akad. Nauk SSSR 9: 250 (1946).

Cyclamen parviflorum Poped. var. **subalpinum** Grey-Wilson, *Cyclamen*: 174 (1997).

Cyclamen pentelici Hildebr. in Bot. Jahrb. 18: no. 44 (1894).

Cyclamen persicum sens. Sibth. & Sm., Fl. Graec. Prod. 1: 128 (1813).

Cyclamen persicum Mill. [var **persicum** forma **persicum**], Gard. Dict. edn. 8, no. 3 (1768).

Cyclamen persicum Mill. subsp. *eupersicum* Knuth in Engl., Pflanzenr. 4, 237: 248 (1905).

Cyclamen persicum Mill. subsp. *mindleri* (Heldr.) Knuth, Pflanzenr. 4, 237: 248 (1905).

Cyclamen persicum Mill. var. **autumnale** Grey-Wilson, *Cyclamen*: 174 (1997).

Cyclamen persicum Mill. var. **persicum** forma **albidum** (Jord.) Grey-Wilson, *Cyclamen*: 174 (1997).

Cyclamen persicum Mill. var. **persicum** forma **puniceum** Grey-Wilson, *Cyclamen*: 174 (1997).

Cyclamen persicum Mill. var. **persicum** forma **roseum** Grey-Wilson nom. provis., *Cyclamen*: 112, 172 (1997).

Cyclamen poli Chiaje, Opusc. Giorn. Med. Nap. 2, fasc 1: 11 (1824).

Cyclamen ponticum (Albov) Poped. in Bot. Zhurn (Moscow & Leningrad) 33: 223 (1948).

Cyclamen pseudibericum Hildebr. [forma **pseudibericum**] in Beih. Centralbl. Bot. 10: 522 (1901).

Cyclamen pseudibericum Hildebr. forma **roseum** Grey-Wilson, *Cyclamen*: 174 (1997).

Cyclamen pseudograecum Hildebr. in Gartenflora 60: 629 (1911).

Cyclamen pseudomaritimum Hildebr. in Gartenflora 57: 293 (1908).

Cyclamen punicum Pomel in Bull. Soc. Bot. France 36: 356 (1889).

Cyclamen purpurascens Mill. [forma **purpurascens**], Gard. Dict. edn. 8, no. 2 (1768).

Cyclamen purpurascens Mill. forma **album** Grey-Wilson, *Cyclamen*: 174 (1997).

Cyclamen purpurascens Mill. subsp. *immaculatum* (Hrabetova) Halda & Soják in Folia Geobot. Phytotax. 3(6): 322 (1971).

Cyclamen purpurascens Mill. subsp. *ponticum* (Albov) Grey-Wilson, Gen. *Cyclamen*: 106 (1988).

Cyclamen pyrolifolium Salisb., Prodr. Stirp. Chap. Allerton: 119 (1796).

Cyclamen rarinaevum Jord. in Jord. & Fourr., Ic. Fl. Eur. 3: 17 (1903).

Cyclamen repandum sens. R.Knuth in Engl., Pflanzenr. 4, 237: 251 (1905).

Cyclamen repandum sens. Texidor, Nuev. Apunt. Fl. Españ.: 23 (1872).

Cyclamen repandum Sm. [subsp. **repandum** var. **repandum** forma **repandum**] in Sibth. & Sm., Fl. Graeca Prodr. 1: 128 (1806).

Cyclamen repandum Sm. subsp. *atlanticum* Maire in Quézel & Santa, Nouv. Fl. Algér.: 724 (1963), without description.

Cyclamen repandum Sm. subsp. *balearicum* (Willk.) Malag., Las Subesp. y Variac. Geogr.: 13 (1973) - [without basionym date].

Cyclamen repandum Sm. subsp. *peloponnesiacum* Grey-Wilson forma *peloponnesiacum* Grey-Wilson, Gen. *Cyclamen*: 60 (1988).

Annex I: All names with author and place of publication - *Cyclamen*
Annexe I: Tous les noms avec les auteurs et le lieu de publication - *Cyclamen*
Anexo I: Todos los nombres, con el autor y el lugar de publicacion - *Cyclamen*

Cyclamen repandum Sm. subsp. *peloponnesiacum* Grey Wilson forma *vividum* Grey-Wilson, Gen. *Cyclamen*: 60 (1988).

Cyclamen repandum Sm. subsp. **peloponnesiacum** var. **peloponnesiacum** (Grey-Wilson) Grey-Wilson, *Cyclamen*: 174 (1997).

Cyclamen repandum Sm. subsp. **peloponnesiacum** Grey-Wilson var. **vividum** (Grey-Wilson) Grey-Wilson, *Cyclamen*: 174 (1997).

Cyclamen repandum Sm. subsp. **repandum** var. **baborense** Debussche & Quézel in Acta Bot. Gallica 144(1): 30 (1997).

Cyclamen repandum Sm. subsp. **repandum** var. **repandum** forma **album** Grey-Wilson, *Cyclamen*: 175 (1997).

Cyclamen repandum Sm. subsp. **rhodense** (Meikle) Grey-Wilson, Gen. *Cyclamen*: 60 (1988).

Cyclamen repandum Sm. var. *creticum* Dörfl. in Ver. K.K. Zool.-Bot. Ges. Wien. 55: 20 (1905).

Cyclamen repandum Sm. var. *rhodense* Meikle in J. Roy. Hort. Soc. 90: 29, pl. 121 (1965).

Cyclamen repandum Sm. var. *stenopetalum* Loret. in Loret & Barrandon, Fl. Montpellier edn. 1, 2: 425 (1827).

Cyclamen retroflexum Moench, Suppl. Meth.: 177 (1802).

Cyclamen rhodium R.Gorer ex O.Schwarz & Lepper in Feddes Repert. Spec. Nov. Regni Veg. 86: 491(1975).

Cyclamen rohlfsianum Asch. in Bull. Herb. Boissier 5: 528 (1897).

Cyclamen romanum Griseb., Spic. Fl. Rumel. 1: 5 (1843).

Cyclamen rotundifolium St.-Lég. in Cariot., Etud. Des Fl. edn. 8, 2: 573 (1899).

Cyclamen sabaudum Jord. in Jord. & Fourr., Ic. Fl. Eur. 3: 20 (1903).

Cyclamen saldense Pomel in Bull. Soc. Bot. France 36: 354 (1889).

Cyclamen × saundersiae Grey-Wilson, *Cyclamen*: 175 (1997), as 'saundersii".

Cyclamen × schwarzii Grey-Wilson, *Cyclamen*: 175 (1997).

Cyclamen somalense Thulin & Warfa in Pl. Syst. Evol., 166(3-4): 249 (1989).

Cyclamen spectabile Jord. in Jord. & Fourr., Ic. Fl. Eur. 3: 16 (1903).

Cyclamen stenopetalum Jord. in Jord. & Fourr., Ic. Fl. Eur. 3: 16 (1903).

Cyclamen subhastatum Rchb., Fl. Germ. Excurs. 1: 407 (1830).

Cyclamen subrotundum Jord. in Jord. & Fourr., Ic. Fl. Eur. 3: 21 (1903).

Cyclamen tauricum hort. Dammann ex Sprenger in Regel, Gartenflora 41: 525 (1892).

Cyclamen trochopteranthum O.Schwarz [forma **trochopteranthum**] in Feddes Repert. Spec. Nov. Regni Veg. 86: 493 (1975).

Cyclamen trochopteranthum O.Schwarz forma **leucanthum** Grey-Wilson, *Cyclamen*: 175 (1997).

Cyclamen tunetanum Jord. in Jord. & Fourr., Ic. Fl. Eur. 3: 23 (1903).

Cyclamen umbratile Jord. in Jord. & Fourr., Ic. Fl. Eur. 3: 18 (1903).

Cyclamen utopicum Hoffmanns., Verz. Pfl.-Kult.: 54 (1824).

Cyclamen variegatum Pohl, Tent. Fl. Bohem. 1: 192 (1810).

Cyclamen velutinum Jord. in Jord. & Fourr., Ic. Fl. Eur. 3: 22, t. 423 (1903).

Cyclamen venustum Jord. in Jord. & Fourr., Ic. Fl. Eur. 3: 22 (1903).

Cyclamen vernale hort. No indication of earliest usage.

Cyclamen vernale sens. O.Schwarz in Feddes Repert. Spec. Nov. Regni Veg. 58: 243 (1955).

Cyclamen vernale Mill., Gard. Dict. edn. 8, no. 4 (1768).

Cyclamen vernum Lobel ex Cambess., Enum. Pl. Balear.: 127 (1827), non Sweet.

Cyclamen vernum Lobel ex Rchb., Fl. Germ. Excurs. 1: 407 (1830).

Cyclamen vernum Sweet, Brit. Flower Gard. 1: t. 9 (1823).

Cyclamen vernum Sweet forma *alpinum* (hort. Dammann ex Sprenger) O.Schwarz in Gartenflora n.s., 1: 20 (1938).

Cyclamen vernum Sweet var. *caucasicum* O.Schwarz in Gartenflora n.s., 1: 20 (1938).

Cyclamen vernum Sweet var. *hiemale* (Hildebr.) O.Schwarz forma *alpinum* (hort. Dammann ex Sprenger) O. Schwarz in Gartenflora n.s., 1: 20 (1938).

Cyclamen vernum Sweet var. *hiemale* (Hildebr.) O.Schwarz forma *pseudocoum* O.Schwarz in Gartenflora n.s., 1: 20 (1938).

Annex I: All names with author and place of publication - *Cyclamen*
Annexe I: Tous les noms avec les auteurs et le lieu de publication - *Cyclamen*
Anexo I: Todos los nombres, con el autor y el lugar de publicacion - *Cyclamen*

Cyclamen × **wellensiekii** Ietsw. in Acta Bot. Neerl. 23(4): 555 (1974).
Cyclamen × **whiteae** Grey-Wilson, *Cyclamen*: 175 (1997), as '**whitei**".
Cyclamen zonale Jord. in Jord. & Fourr., Ic. Fl. Eur. 3: 15, t. 401 (1903).
Cyclaminos graeca (Link) Heldr. in Bull. Herb. Boiss. 6: 386 (1898).
Cyclaminos miliarakesii Heldr., Herb. norm. graec. no 1575 (1900).
Cyclaminos mindleri Heldr. in Bull. Herb. Boiss. 6: 386 (1898).
Cyclaminum vernum Bubani., Fl. Pyren. 1: 229 (1897).
Cyclaminus coa (Mill.) Asch. in Ber. Deutsch. Bot. Ges. 10: 235 (1892).
Cyclaminus europaea (L.) Asch. in Ber. Deutsch. Bot. Ges. 10: 235 (1892).
Cyclaminus europaeus [*sic*] (L.) Scop., Fl. Carniol edn. 2, 1: 136 (1772).
Cyclaminus graeca (Link) Asch. in Ber. Deutsch. Bot. Ges. 10: 235 (1892).
Cyclaminus neopolitana (Ten.) Asch. in Ber. Deutsch. Bot. Ges. 10: 235 (1892).
Cyclaminus persica (Mill.) Asch. in Ber. Deutsch. Bot. Ges. 10: 235 (1892).
Cyclaminus repanda (Sm.) Asch. in Ber. Deutsch. Bot. Ges. 10: 235 (1892).

Annex I: All names with author and place of publication - *Galanthus*
Annexe I: Tous les noms avec les auteurs et le lieu de publication - *Galanthus*
Anexo I: Todos los nombres, con el autor y el lugar de publicacion - *Galanthus*

GALANTHUS

Chianthemum elwesii (Hook.f.) Kuntze, Rev. Gen. Pl. 2: 703 (1891). †
Chianthemum graecum (Orph. ex Boiss.) Kuntze, Rev. Gen. Pl. 2: 703 (1891). †
Chianthemum nivale (L.) Kuntze, Rev. Gen. Pl. 2: 703 (1891). †
Chianthemum olgae (Orph.) Kuntze, Rev. Gen. Pl. 2: 703 (1891). †
Chianthemum plicatum (M.Bieb.) Kuntze, Rev. Gen. Pl. 2: 703 (1891). †
Galanthus alexandrii Porcius, Anal. Acad. Române 14: 274 (1893).
Galanthus allenii Baker in, Gard. Chron. ser. 3, 9: 298 (1891), as '*alleni*'.
Galanthus × **allenii** (?*Galanthus alpinus* Sosn. × *Galanthus woronowii* Losinsk.).
Galanthus alpinus Sosn. [var. **alpinus**] in Vestn. Tiflissk. Bot. Sada 19: 26 (1911).
Galanthus alpinus Sosn. var. **bortkewitschianus** (Koss) A.P.Davis in A.P.Davis, Mordak & Jury in Kew Bull. 51(4): 750 (1996).
Galanthus angustifolius Koss in Bot. Mater. Gerb. Inst. Kom. a Akad. Nauk SSSR 14: 134, fig. 3 (1951).
Galanthus atkinsi hort. Barr (1875) - possibly offered in a Barr catologue under this name, see J.Allen in Garden (London) 40: 272 (1891). †
Galanthus bortkewitschianus Koss in Bot. Mater. Gerb. Inst. Kom. a Akad. Nauk SSSR 14: 130, fig. 1 (1951).
Galanthus bulgaricus Velen., Fl. Bulg.: 539 (1891). †
Galanthus byzantinus Baker in Gard. Chron. ser. 3, 13: 226 (1893).
Galanthus byzantinus Baker subsp. *brauneri* N.Zeybek in Doga. Tu. J. Botany 12(1): 101 (1988), as "*braunerii*"
Galanthus byzantinus Baker subsp. *saueri* N.Zeybek in Doga. Tu. J. Botany 12(1): 101 (1988), as "*sauerii*".
Galanthus byzantinus Baker subsp. *tughrulii* N.Zeybek in Doga. Tu. J. Botany 12(1): 100 (1988).
Galanthus cabardensis Koss in Bot. Mater. Gerb. Inst. Kom. a Akad. Nauk SSSR 14: 133, fig. 2 (1951).
Galanthus caspius (Rupr.) Grossh., Fl. Caucas. edn. 2, 2: 193, t. 24, fig. 6, map 223 (1940). †
Galanthus caucasicus (Baker) Burb. in J. Roy. Hort. Soc. 13(2): 200 (1891). †
Galanthus caucasicus (Baker) Grossh. in Grossh. & Schischk., Pl. Orient. Exsicc. fasc. 1: 4, no. 6 (1924).
Galanthus caucasicus (Baker) Grossh. var. *hiemalis* Stern in J. Roy. Hort. Soc. 86(7): 324 (1961). †
Galanthus caucasicus hort. - sens. auct.: Stern, Snowdr. & Snowfl.: 137, fig. 14 (1956), non (Baker) Grossh.
Galanthus cilicicus auct. non Baker: Artjush. in Bot. Zhurn. (Moscow & Leningrad) 51(10): 1449 (1966).
Galanthus cilicicus auct. non Baker, pro parte: Artjush. in Ann. Mus. Goulandris 2: 18 (1974).
Galanthus cilicicus Baker in Gard. Chron. ser. 3, 21: 214 (1897).
Galanthus cilicicus Baker subsp. *caucasicus* O.Schwarz in Bull. Alp. Gard. Soc. Gr. Brit. 31: 134 (1963). †
Galanthus clusii Fisch. sens. Steud., Nomencl. Bot. edn. 2, 1: 653 (1840), in syn. †
Galanthus corcyrensis (Beck) Stern, Snowdr. & Snowfl.: 34, fig. 7 (p. 35) (1956).
Galanthus corcyrensis Burb. in J. Roy. Hort. Soc. 13(2): 201 (1891). †
Galanthus corcyrensis J.Allen in Garden (London) 29: 75 (1886). †
Galanthus corcyrensis Leichtlin ex Correvon in Le Jardin 2: 139 (1888). †
Galanthus corcynensis [sic] T.Shortt in Gard. Chron. n.s., 20: 728 (1883). †
Galanthus corcyrensis (*praecox*) [sic] hort. ex Baker, Handb. Amaryll.: 17 (1888). †
Galanthus elsae Burb. in Gard. Chron. ser. 3, 7: 268 (1890). †
Galanthus elsae Burb. in Garden (London) 39: 243 (1891).
Galanthus elsae Ewbank in Garden (London) 39: 273 (1891). †
Galanthus elsae J.Allen in J. Roy. Hort. Soc. 13(2): 181 (1891). †
Galanthus elwesii Hook.f. in Bot. Mag. 101: t. 6166 (1875).
Galanthus elwesii Hook.f. subsp. *akmanii* N.Zeybek in Doga. Tu. J. Botany 12(1): 99 (1988).

Annex I: All names with author and place of publication - *Galanthus*
Annexe I: Tous les noms avec les auteurs et le lieu de publication - *Galanthus*
Anexo I: Todos los nombres, con el autor y el lugar de publicacion - *Galanthus*

Galanthus elwesii Hook.f. subsp. *baytopii* (N.Zeybek) N.Zeybek & E.Sauer, Türk. Kardelenleri 1/Beitr. Kennt. Türk. Schneeglöckhen 1: 60, fig. 17 (1995).
Galanthus elwesii Hook.f. subsp. *melihae* N.Zeybek in Doga. Tu. J. Botany 12(1): 98 (1988).
Galanthus elwesii Hook.f. subsp. *minor* D.A.Webb in Bot. J. Linn. Soc. 76(4): 312 (1978).
Galanthus elwesii Hook.f. subsp. *tuebitaki* N.Zeybek in Doga. Tu. J. Botany 12(1): 98 (1988).
Galanthus elwesii Hook.f. subsp. *wagenitzii* N.Zeybek in Doga. Tu. J. Botany 12(1): 98 (1988).
Galanthus elwesii Hook.f. subsp. *yayintaschii* N.Zeybek in Doga. Tu. J. Botany 12(1): 98 (1988).
Galanthus elwesii Hook.f. [var.] *globosus* Ewbank in The Garden (London) 39: 272 (1891).†
Galanthus elwesii Hook.f. var. *maximus* (Velen.) Beck in Wiener Ill. Gart.-Zeitung 19: 55 fig. 2,14 (1894).
Galanthus elwesii Hook.f. var. *monostictus* P.D.Sell in P.D.Sell & G.Murrell, Fl. Gr. Brit. & Irel. 5: 363(1996).
Galanthus elwesii Hook.f. var. *platyphyllus* Kamari in Bot. Jahrb. Syst. 122 (1): 122 (1982).
Galanthus elwesii Hook.f. var. *reflexus* (Herb. ex Lindl.) Beck in Wiener Ill. Gart.-Zeitung 19: 55 (1894).
Galanthus elwesii Hook.f. var. *robustus* Baker in Gard. Chron. ser. 3, 13: 226 (1893).
Galanthus elwesii Hook.f. var. *stenophyllus* Kamari in Bot. Jahrb. Syst. 122 (1): 122 (1982).
Galanthus elwesii Hook.f. var. *whittallii* (hort.) W.Irving in Garden (London) 59: 262 (1901).
Galanthus elwesii Hook.f. var. *whittallii* Moon in Garden (London) 57: 44 (1900). †
Galanthus elwesii Hook.f. var. *whittallii* S.Arn. in Gard. Chron. ser. 3, 24(2): 466 (1898). †
Galanthus fosteri Baker in Gard. Chron. ser. 3, 5: 458 (1889).
Galanthus fosteri Baker var. *antepensis* N.Zeybek & E.Sauer, Türk. Kardelenleri 1/Beitr. Kennt. Türk. Schneeglöckhen 1: 72, fig. 25 (1995) †.
Galanthus glaucescens A.P.Khokhr. in Byull. Glavn. Bot. Sada 62: 62 (1966).
Galanthus globosus Burb. in J. Roy. Hort. Soc. 13: 203, fig. 28 (1891).
Galanthus globosus Wilks in Garden (London) 31: 393 (1887). †
Galanthus gracilis Čelak. in Sber. K. böhm. Ges. Wiss. Cl. 1891, t. 9: 195 (1891).
Galanthus gracilis Čelak. subsp. *baytopii* N.Zeybek in Doga. Tu. J. Botany 12(1): 96 (1988).
Galanthus graecus auct. non Orph. ex Boiss., pro parte excl. *G. gracilis* Čelak.: Stern, Snowdr. & Snowfl.: 40 (1956); Artjush. in Pl. Life 25(2-4): 146 (1969); Delip. in Izv. Bot. Inst. 21: 163 (1971); Artjush. in Ann. Mus. Goulandris 2: 14 (1974).
Galanthus graecus auct. non Orph. ex Boiss.: P.H.Davis in Kew Bull. no. 1: 113 (1949).
Galanthus graecus Orph. ex Boiss., Fl. Orient. 5: 145 (1882).
Galanthus graecus Orph. ex Boiss. forma *gracilis* (Čelak.) Zahar. in Savul. & Nyár., Fl. Republ. Social. Romania 11: 413 (1966).
Galanthus graecus Orph. ex Boiss. forma *maximus* (Velen.) Zahar. in Savul. & Nyár., Fl. Republ. Social. Romania 11: 413 (1966).
Galanthus graecus Orph. var. *maximus* (Velen.) Hayek in Fedde, Prod. Fl. Penins. Balcan. 30: 3, 102 (1932).
Galanthus × grandiflorus Baker in Gard. Chron. ser. 3, 13: 656 (1893).
Galanthus grandis Burb. in J. Roy. Hort. Soc. 13(2): 203 (1891). †
Galanthus ikariae auct. non Baker, pro parte, in syn.: Artjush. in Bot. Zhurn. (Moscow & Leningrad) 50(10): 1445 (1965); Artjush. in Bot. Zhurn. (Moscow & Leningrad) 51(10): 1448, fig. 5 (1966); Artjush. in Daffodil Tulip Year Book: 70, fig. 5, map 20 (1967); Artjush. in Pl. Life 25(2-4): 148, fig. 30 (1969); C.D.Brickell in P.H.Davis et al. [eds.], Fl. Turkey 8: 372, map 55 (1984); Brickell in Walters et al. [eds.], Eur. Gard. Fl. 1: 319 (1986); Mordak in Spisok rast. Gerb. flory SSSR 27(92): No. 7107 (1990).
Galanthus ikariae Baker in Gard. Chron. ser. 3, 13: 506 (1893).
Galanthus ikariae auct. non Baker: T.Baytop & B.Mathew, Bulb. Pl. Turkey: 22 (1984).
Galanthus ikariae Baker subsp. *latifolius* Stern, Snowdr. & Snowfl.: 50, fig. 13 (p. 51) (1947) - (name includes more than one taxon - pro parte).
Galanthus ikariae Baker subsp. *snogerupii* Kamari in Bot. Chron. 1(2): 76 (1981) †; Kamari in Bot. Jahrb. Syst. 103(1): 124, 126; fig. 8 (1982), descr.
Galanthus imperati Bertol., Fl. Ital. 4: 5 (1839).
Galanthus imperati Bertol. forma *australis* Zodda, Fl. Ital. Exsicc. num. 762 (1904), in shed.

Annex I: All names with author and place of publication - *Galanthus*
Annexe I: Tous les noms avec les auteurs et le lieu de publication - *Galanthus*
Anexo I: Todos los nombres, con el autor y el lugar de publicacion - *Galanthus*

Galanthus kemulariae Kuth. in Zametki Sist. Geogr. Rast. 23: 128 (1963).

Galanthus ketzkhovelii Kem.-Nath. in Trudy Tbilissk Bot. Inst. ser. 2, 11: 181 (1947).

Galanthus koenenianus Lobin, C.D.Brickell & A.P.Davis in Kew Bull. 48(1): 161 (1993).

Galanthus krasnovii A.P.Khokhr. in Bjull. Moskovsk. Obsc. Isp. Prir. Otd. Biol. 68 (4): 140 (1963).

Galanthus krasnovii A.P.Khokhr. subsp. *maculatus* A.P.Khokhr. in Bjull. Glavn. Bot. Sada 62: 60 (1966).

Galanthus lagodechianus Kem.-Nath. in Makaschv., Sosn. & Kharadze (eds.), Fl. Gruzii [Fl. Georgiae], 2: 526 (1941) †; Kem.-Nath. in Zametki Sist. Geogr. Rast. 13: 6 (1947), descr.

Galanthus latifolius auct. non Rupr.: Masters in Gard. Chron. n.s., 15: 404 (1881); Stern & Gilmour in Bot. Mag. 164: t. 9669 (1946).

Galanthus latifolius Rupr. in Gartenflora 17: 130, t. 578 (1868). †

Galanthus latifolius Salisb., Gen. Pl.: 95 (1866). †

Galanthus latifolius Rupr. forma *allenii* (Baker) Beck in Wiener Ill. Gart.-Zeitung 19: 56, fig. 2,16 (1894).

Galanthus latifolius Rupr. forma *fosteri* (Baker) Beck in Wiener Ill. Gart.-Zeitung 19: 57, fig. 2,17 (1894).

Galanthus latifolius Rupr. forma *typicus* Beck in Wiener Ill. Gart.-Zeitung 19: 49 (1894).

Galanthus latifolius Rupr. forma *typicus.* Gottl.-Tann in Abh. K. K. Zool.-Bot. Ges. Wien 2(4): 42 (1904) †.

Galanthus latifolius Rupr. [var.] *rizaensis* [*sic*] Anon. in Proc. J. Roy. Hort. Soc. 60: xxv (1935). †

Galanthus latifolius Rupr. var. *rizehensis* Stern & Gilmour in Bot. Mag. 164: t. 9669 (1946).†

Galanthus maximus Velen., Fl. Bulg.: 540 (1891).

Galanthus × *maximus* Baker (n. sp. or hybrid ?) [*sic*] in Gard. Chron. 13: 254 (1893), non Velen. †

Galanthus melihae (N.Zeybek) N.Zeybek & E.Sauer, Türk. Kardelenleri 1/Beitr. Kennt. Türk. Schneeglöckchen 1: 54, figs. 12 [bottom left], 13, 14 (1995).

Galanthus montana Schur, Enum. Pl. Transsilv.: 658 (1866).

Galanthus nivalis L. 'Flore Pleno' hort. (earliest usage not known).

Galanthus nivalis L., Sp. Pl. 1: 288 (1753).

Galanthus nivalis L. forma *octobrinus* [*sic*] Voss in Siebert & Voss, Vilm. Blumengärt. edn. 3, 1: 1006 (1895).

Galanthus nivalis L. forma *pictus* Maly in Verh. Zool.-Bot. Ges. Wien 54: 302 (1904).

Galanthus nivalis L. forma *pleniflorus* P.D.Sell in P.D.Sell & G.Murrell, Fl. Gr. Brit. & Irel. 5: 363 (1996).

Galanthus nivalis L. subsp. *allenii* (Baker) Gottl.-Tann. in Abh. K. K. Zool.-Bot. Ges. Wien 2(4): 37 (1904). †

Galanthus nivalis L. subsp. *angustifolius* (Koss) Artjush., Amaryllidaceae SSSR: 75, [fig. 44 = *G. nivalis*] (1970).

Galanthus nivalis L. subsp. *byzantinus* (Baker) Gottl.-Tann. in Abh. K. K. Zool.-Bot. Ges. Wien 2(4): 41 (1904).

Galanthus nivalis L. subsp. *caucasicus* Baker in Gard. Chron. ser. 3, 13: 313 (1887).

Galanthus nivalis L. subsp. *cilicicus* auct. non (Baker) Gottl.-Tann. pro parte: Kamari in Bot. Jahrb. Syst. 103(1): 111 (1982).

Galanthus nivalis L. subsp. *cilicicus* (Baker) Gottl.-Tann. in Abh. K. K. Zool.-Bot. Ges. Wien 2(4): 33 (1904).

Galanthus nivalis L. subsp. *elwesii* (Hook.f.) Gottl.-Tann. in Abh. K. K. Zool.-Bot. Ges. Wien 2(4): 39 (1904).

Galanthus nivalis L. subsp. *graecus* (Orph. ex Boiss.) Gottl.-Tann. in Abh. K. K. Zool.-Bot. Ges. Wien 2(4): 40 (1904), pro parte excl. *G. gracilis* Čelak.

Galanthus nivalis L. subsp. *humboldtii* N.Zeybek in Doga. Tu. J. Botany 12(1): 97 (1988).

Galanthus nivalis L. subsp. *imperati* (Bertol.) Baker, Handb. Amaryll.: 17 (1888).

Galanthus nivalis L. subsp. *plicatus* (M.Bieb.) Gottl.-Tann. in Abh. K. K. Zool.-Bot. Ges. Wien 2(4): 35 (1904).

Galanthus nivalis L. subsp. *reginae-olgae* (Orph.) Gottl.-Tann. in Abh. K. K. Zool.-Bot. Ges. Wien 2(4): 32 (1904).

Annex I: All names with author and place of publication - *Galanthus*
Annexe I: Tous les noms avec les auteurs et le lieu de publication - *Galanthus*
Anexo I: Todos los nombres, con el autor y el lugar de publicacion - *Galanthus*

Galanthus nivalis L. subsp. *subplicatus* (N.Zeybek) N.Zeybek & E.Sauer, Türk. Kardelenleri 1/Beitr. Kennt. Türk. Schneeglöckhen 1: 47, fig. 9 (1995).
Galanthus nivalis L. [var.] *atkinsii* J.Allen in Garden (London) 40: 272 (1891). †
Galanthus nivalis L. var. *atkinsii* Mallett in Garden (London) 67: 87 (1905).
Galanthus nivalis L. var. *carpaticus* Fodor in Ukrajins'k. Bot. Zhurn. 40(5): 32, fig. 1 (1983).
Galanthus nivalis L. var. *caspius* Rupr. in Gartenflora 17: 132 (1868).
Galanthus nivalis L. var. *caucasicus* (Baker) Beck in Wiener Ill. Gart.-Zeitung: 19 (1894).
Galanthus nivalis L. var. *caucasicus* (Baker) Fomin in Fomin & Woronow, Opred. Rast. Caucas. Krym. 1: 280 (1909). †
Galanthus nivalis L. var. *caucasicus* (Baker) J. Phillippow in Kuzn. et al., Fl. Cauc. Crit. 2.5: 5 (1916). †
Galanthus nivalis L. var. *corcyrensis* (Beck) Halácsy, Consp. Fl. Graec. 3: 206 (1904).
Galanthus nivalis L. [var.] *corcyrensis* hort. ex Leichtlin in Gard. & Forest 1: 499 (1888). †
Galanthus [*nivalis* L.] var. *elsae* Mallett in Garden (London) 67: 87 (1905). †
Galanthus nivalis L. var. *europaeus* forma *corcyrensis* hort. ex Beck in Wiener Ill. Gart.-Zeitung 19: 51, fig. 1,6 (1894).
Galanthus nivalis L. var. *europaeus* forma *hololeucus* (Čelak.) Beck in Wiener Ill. Gart.-Zeitung 19: 50 (1894).
Galanthus nivalis L. var. *europaeus* forma *hortensis* (Herb.) Beck in Wiener Ill. Gart.-Zeitung 19: 50 (1894).
Galanthus nivalis L. var. *europaeus* forma *olgae* (Orph.) Beck in Wiener Ill. Gart.-Zeitung 19: 51, fig. 1,5 (1894).
Galanthus nivalis L. var. *europaeus* forma *scharloki* [*sic*] (Casp.) Beck in Wiener Ill. Gart.-Zeitung 19: 52, fig. 1,7 (1894).
Galanthus nivalis L. var. *grandior* Schult. & Schult.f., Syst. Veg. 7(2): 781 (1830).
Galanthus nivalis L. var. *hololeucus* Čelak. in Abh. Königl. Böhm. Ges. Wiss.: 198 (1891).
Galanthus nivalis L. var. *hortensis* Herb., Amaryll.: 330 (1837).
Galanthus nivalis L. var. *imperati* Mallett in Garden (London) 67: 87 (1905). †
Galanthus nivalis L. var. *major* Redouté (in descr., Redouté, Liliac.: t. 200) ex Rupr. in Regel, Gartenflora 17: 131 (1868) [! descr. t. 200 = *G. nivalis* L.]. †
Galanthus nivalis L. var. *major* sens. Fiori, Nouv. Fl. Analit. Italia 1: 286 (1923), non Redouté.
Galanthus nivalis L. var. *majus* [*sic*] Ten., Fl. Napol. 1: 140 (1811–1815).
Galanthus nivalis L. var. *maximus* (Velen.) Stoj. & Stevanov, Fl. Bulg. 4: 257 (1923).
Galanthus nivalis L. var. *minus* Ten., Fl. Napol. 1: 140 (1811–1815).
Galanthus nivalis L. var. *montanus* (Schur) Rouy, Fl. France 13: 21 (1912).
Galanthus nivalis L. var. *octobrensis* Mallett in Garden (London) 67: 87 (1905). †
Galanthus nivalis L. var. *praecox* Mallett in Garden (London) 67: 87 (1905). †
Galanthus nivalis L. var. *rachelae* Mallett in Garden (London) 67: 87 (1905). †
Galanthus nivalis L. var. *redoutei* Rupr. ex Regel in Gartenflora 12: 177, pl. 400, fig. 2 (1863).
Galanthus nivalis L. var. *reginae-olgae* (Orph.) Fiori, Fl. Anal. Italia 1: 286 (1923).
Galanthus nivalis L. var. *scharlockii* Casp. in Schriften Königl. Phys.-Ökon. Ges. Königsberg. 1-72: 18 (1868).
Galanthus nivalis L. var. *shaylockii* [*sic*] Harpur-Crewe in Gard. Chron. n.s., 11: 342, fig. 48 (1879). †
Galanthus nivalis L. var. *typicus* Rouy, Fl. France 13: 20 (1912). †
Galanthus nivalis sens. Ledeb., Fl. Rossica 4: 113 (1853), pro parte.
Galanthus octobrensis Burb. in Gard. Chron. ser. 3, 7: 268 (1890). †
Galanthus octobrensis Burb. in Garden (London) 39: 243 (1891). †
Galanthus octobrensis Ewbank in Garden (London) 39: 272 (1891). †
Galanthus octobrensis hort. ex Baker, Handb. Amaryll.: 17 (1888). †
Galanthus octrobrensis hort. ex Burb. in J. Roy. Hort. Soc. 13(2): 206 (1891). †
Galanthus octobrensis J.Allen in Garden (London) 29: 75 (1886). †
Galanthus octobrensis J.Allen in J. Roy. Hort. Soc. 13(2): 179 (1891). †
Galanthus octobrensis Leichtlin ex Correvon in Le Jardin 2: 139 (1888). †
Galanthus octobrensis T.Shortt in Gard. Chron. n.s., 20: 728 (1883). †
Galanthus olgae Orph. ex Boiss., Fl. Orient. 5: 146 (1882). †

Annex I: All names with author and place of publication - *Galanthus*
Annexe I: Tous les noms avec les auteurs et le lieu de publication - *Galanthus*
Anexo I: Todos los nombres, con el autor y el lugar de publicacion - *Galanthus*

Galanthus olgae reginae hort. ex Leichtlin in Gard. & Forest 1: 499 (1888). †
Galanthus perryi hort. Ware ex Baker in Gard. Chron. ser. 3, 13: 258 (1893).
Galanthus peshmenii A.P.Davis & C.D.Brickell in The New Plantsman 1: 14 (1994).
Galanthus platyphyllus Traub & Moldenke in Herbertia 14: 110 (1948).
Galanthus plicatus auct. non M. Bieb.: Hohen., Enum. Pl. Elisazethp.: 228 (1883).
Galanthus plicatus sens. Guss., Plantae Rariores: 140 (1826), non Salisb., non M.Bieb., non Hohen.
Galanthus plicatus M.Bieb. [subsp. **plicatus**], Fl. Taur.-Caucas. 3: 255 (1819).
Galanthus plicatus M.Bieb. subsp. **byzantinus** (Baker) D.A.Webb in Bot. J. Linn. Soc., 76(4): 310 (1978).
Galanthus plicatus M.Bieb. subsp. *gueneri* N.Zeybek in Doga. Tu. J. Botany 12(1): 99 (1988), as "*guenerii*".
Galanthus plicatus M.Bieb. subsp. *karamanoghluensis* N.Zeybek in Doga. Tu. J. Botany 12(1): 100 (1988).
Galanthus plicatus M. Bieb. subsp. *plicatus* var. *viridifolius* P.D.Sell in P.D.Sell & G.Murrell, Fl. Gr. Brit. & Irel. 5: 363 (1996).
Galanthus plicatus M.Bieb. subsp. *subplicatus* N.Zeybek in Doga. Tu. J. Botany 12(1): 99 (1988).
Galanthus plicatus M.Bieb. subsp. *vardarii* N.Zeybek in Doga. Tu. J. Botany 12(1): 100 (1988).
Galanthus plicatus M.Bieb. var. *byzantinus* (Baker) Beck in Wiener Ill. Gart.-Zeitung 19: 57 (1894).
Galanthus plicatus M.Bieb. var. *genuinus* forma *excelsior* Beck in Wiener Ill. Gart.-Zeitung 19: 57, fig. 2,19 (1894). †
Galanthus plicatus M.Bieb. var. *genuinus* forma *maximus* Beck in Wiener Ill. Gart.-Zeitung19: 57 (1894). †
Galanthus plicatus M.Bieb. var. *genuinus* forma *typicus* Beck in Wiener Ill. Gart.-Zeitung 19: 57, fig. 2,18 (1894). †
Galanthus praecox Burb. in Garden (London) 39: 243 (1891).
Galanthus praecox Burb. in J. Roy. Hort. Soc. 13(2): 207 (1891). †
Galanthus praecox J.Allen in Garden (London) 29: 75 (1886). †
Galanthus rachelae Burb. in Gard. Chron. ser. 3, 7: 268 (1890). †
Galanthus rachelae Burb. in Garden (London) 39: 243 (1891).
Galanthus rachelae Burb. in J. Roy. Hort. Soc. 13(2): 207 (1891). †
Galanthus rachelae Ewbank in Garden (London) 39: 272 (1891). †
Galanthus rachelae J.Allen in J. Roy. Hort. Soc. 13(2): 180 (1891). †
Galanthus redoutei (Rupr. ex Regel) Regel, Gartenflora 23: 202 (1874). †
Galanthus reflexus Herb. ex Lindl. in Edward's Bot. Reg. 31: 35, misc. 44 (1845).
Galanthus reflexus auct. non Herb. ex Lindl.: Baker, Handb. Amaryll.: 17 (1888); Harpur-Crewe in Gard. Chron. n.s., 11: 237 (1879); Burb. in J. Roy. Hort. Soc. 13(2): 208 (1891); Stern, Snowdr. & Snowfl.: 75 (1956). †
Galanthus reginae-olgae auct. non Orph., pro parte: C.D.Brickell in P.H.Davis et al. [eds.], Fl. Turkey 8: 367, map 54 (1984).
Galanthus reginae-olgae Orph. [subsp. **reginae-olgae**] in Atti Congr. Intern. Botan. Firenze: 214 (1876).
Galanthus reginae-olgae Orph. subsp. *corcyrensis* (Beck) Kamari in Bot. Chron. 1(2): 68 (1981) †; Kamari in Bot. Jahrb. Syst. 103(1): 115 (1982), descr.
Galanthus reginae-olgae Orph. subsp. **vernalis** Kamari in Bot. Chron. 1(2): 68 (1981) †; Kamari in Bot. Jahrb. Syst. 103(1): 116 (1982), descr.
Galanthus rizehensis Stern, Snowdr. & Snowfl.: 37 (1956).
Galanthus schaoricus Kem.-Nath. in Makashv. & Sosn., Fl. Georgia 2: 526 (1941) †; Kem.-Nath. in Zametki Sist. Geogr. Rast. 13: 6 (1947), descr.
Galanthus sharlockii (Casp.) Baker, Handb. Amaryll.: 17 (1888).
Galanthus shaylockii [*sic*] J.Allen in Garden (London) 29: 75 (1886). †
Galanthus transcaucasicus Fomin in Fomin & Woronow, Opred. Rast. Caucas. Krym. 1: 281 (1909).
Galanthus valentinae Panjut. (in sched. (1938), leg. 1913) ex Grossh., Fl. Caucas. edn. 2, 2: 194, map 212 (1940). †

Annex I: All names with author and place of publication - *Galanthus*
Annexe I: Tous les noms avec les auteurs et le lieu de publication - *Galanthus*
Anexo I: Todos los nombres, con el autor y el lugar de publicacion - *Galanthus*

Galanthus woronowii Losinsk. in Kom., Fl. URSS 4: 749 (1935).

† Names not validly published or names invalid (only given for *Galanthus* and *Sternbergia*).

† Noms non publiés validement ou noms non valides (donnés seulement pour *Galanthus* et *Sternbergia*)

† Nombres publicados que carecen de validez o nombres inválidos (indicados únicamente para *Galanthus* y *Sternbergia*)

Annex I: All names with author and place of publication - *Sternbergia*
Annexe I: Tous les noms avec les auteurs et le lieu de publication - *Sternbergia*
Anexo I: Todos los nombres, con el autor y el lugar de publicacion - *Sternbergia*

STERNBERGIA

Amaryllis aetnensis Raf., Caratt.: 84, t. 18 f. 3 (1810).
Amaryllis citrina Sibth. & Sm., Fl. Graec. 4: 11, t. 311 (1823).
Amaryllis clusiana Ker Gawl., in Bot. Mag. 27: sub t. 1089 (1808).
Amaryllis colchiciflora (Waldst. & Kit.) Ker Gawl. in Bot. Mag. 27: sub. t. 1089 (1808).
Amaryllis lutea L., Sp. Pl. 2: 292 (1753).
Amaryllis lutea M.Bieb., Fl. Taur.-Cauc. 3: 215 (1819).
Amaryllis vernalis Mill., Gard. Dict. cdn. 8, no. 10 (1768).
Oporanthus colchiciflorus (Waldst. & Kit.) Herb., Appendix: 38 (1821).
Oporanthus fischerianus Herb., Amaryll.: 412 (1837).
Oporanthus luteus (L.) Herb., Appendix: 38 (1821).
Oporanthus luteus (L.) Herb. var. *angustifolia* Herb., Appendix: 38 (1821).†
Oporanthus luteus (L.) Herb. var. *latifolia* Herb., Appendix: 38 (1821).†
Sternbergia aetnensis (Raf.) Guss., Fl. Sic. Prod. 1: 385 (1827).
Sternbergia alexandrae Sosn. in Trudy Azerbajdzansk. Otd. Zakavkazsk. Fil. Akad. Nauk SSSR 2: 269 (1936).
Sternbergia americana Hoffmanns., Verz. Pfl.-Kult.: 197 (1824).
Sternbergia aurantiaca Dinsm. in Post, Fl. Syria edn. 2, 2: 607 (1934).
Sternbergia candida B.Mathew & T.Baytop in The Garden 104(7): 302 (1979).
Sternbergia caucasica Willd. in Ges. Naturf. Freunde Berlin Neue Schriften 2: 27 (1808).
Sternbergia citrina (Sibth. & Sm.) Schult. & Schult.f., Syst. Veg. 7: 795 (1830).
Sternbergia clusiana (Ker Gawl.) Spreng., Syst. Veg. 2: 57 (1825).
Sternbergia colchiciflora Waldst. & Kit., Desc. Ic. Pl. Rar. Hung. 2: 172, t. 159 (1803-4).
Sternbergia colchiciflora Waldst. & Kit. var. *aetnensis* (Raf.) Rouy in Bull. Soc. Bot. France 31: 182 (1884).
Sternbergia colchiciflora Waldst. & Kit. var. *alexandrae* (Sosn.) Artjush., Amaryllidaceae SSSR: 95 (1970).
Sternbergia colchiciflora Waldst. & Kit. var. *dalmatica* Herb., Amaryll.: 413, t. 47, fig. 2 (1837).
Sternbergia dalmatica (Herb.) Herb., Amaryll.: 187 & 413, t. 47 fig. 2 (1837), ? comb. provis., Baker.
Sternbergia exigua Schult. & Schult.f., Syst. Veg. 7: 795 (1830).
Sternbergia exscapa Guss., Fl. Sicul. Syn. 1: 384 (1843).
Sternbergia fischeriana (Herb.) M.Roem., Fam. Nat. Syn. Monogr. 4: 46 (1847).
Sternbergia fischeriana (Herb.) M.Roem. forma *hissarica* Kapinos, Biol. Zakon. razv. lukov. i Klubnelukov. rast. na Apsherone: 77 (1965).
Sternbergia fischeriana (Herb.) M.Roem. subsp. *hissarica* (Kapinos) Artjush., Amaryllidaceae SSSR: 97 (1970).
Sternbergia fischeriana (Herb.) Rupr. in Regel, Gartenflora 17: 100, t. 576 (1868).†
Sternbergia grandiflora Boiss., Plant. Cilic. Exsicc.: No. 344, nom. in sched. (based on *Kotschy* 344); Baker in Bot. Mag. 122: sub t. 7459 (1895), in syn.
Sternbergia greuteriana Kamari & R.Artlelari in Willdenowia 19(2): 371 (1990).
Sternbergia latifolia Boiss. & Hausskn., Pl. Orient. Exsicc. 1867, nom. in sched. (based on *Haussknecht* s.n., 1867).
Sternbergia lutea Ker Gawl. ex Schult. & Schult.f., Syst. Veg. 7: 795 (1830).†
Sternbergia lutea (L.) Ker Gawl. ex Spreng., Syst. Veg. 2: 57 (1825).
Sternbergia lutea (L.) Spreng. subsp. *sicula* (Tineo ex Guss.) D.A.Webb in Bot. J. Linn. Soc. 76(4): 358 (1978).
Sternbergia lutea Orph., Consp. Fl. Graec.: 142 (1850).
Sternbergia lutea (L.) Spreng. var. *graeca* Rchb., Icon. Fl. Germ. Helv. 9: t. 372, fig. 828 (1847).†
Sternbergia lutea (L.) Spreng. var. *sicula* (Tineo ex Guss.) Tornab., Fl. Aetnea 4: 95 (1892).
Sternbergia macrantha J.Gay ex Baker, Handb. Amaryll.: 29 (1888).
Sternbergia pulchella Boiss. & Blanche, Diagn. Pl. Orient. ser. 2, 3(4): 97 (1859).

Annex I: All names with author and place of publication - *Sternbergia*
Annexe I: Tous les noms avec les auteurs et le lieu de publication - *Sternbergia*
Anexo I: Todos los nombres, con el autor y el lugar de publicacion - *Sternbergia*

Sternbergia schubertii Schenk, Pl. Sp. Aegypt. Arab. et Syriam: 11 (1840).
Sternbergia sicula Tineo ex Guss., Fl. Sic. Syn. 2(2): 811(1845).
Sternbergia spaffordiana Dinsm. in Feddes, Rep. 24: 302 (1928).
Sternbergia stipitata Boiss. & Hausskn. ex Boiss., Fl. Orient. 5: 148 (1882).
Sternbergia vernalis (Mill.) R.Gorer & J.H.Harvey in The Plantsman 10(4): 204 (1989).

† Names not validly published or names invalid (only given for *Galanthus* and *Sternbergia*).

† Noms non publiés validement ou noms non valides (donnés seulement pour *Galanthus* et *Sternbergia*)

† Nombres publicados que carecen de validez o nombres inválidos (indicados únicamente para *Galanthus* y *Sternbergia*)

Annex II: Identification keys
Annexe II: Cles de determination
Anexo II: Claves para la identificacion

ANNEX II: IDENTIFICATION KEYS
For the genera:

Cyclamen, *Galanthus* and *Sternbergia*

ANNEXE II: CLES DE DETERMINATION
Pour les genre:

Cyclamen, *Galanthus* et *Sternbergia*

ANEXO II: CLAVES PARA LA IDENTIFICACION
Para los géneros:

Cyclamen, *Galanthus* y *Sternbergia*

IDENTIFICATION KEYS FOR CYCLAMEN, GALANTHUS AND STERNBERGIA

KEY TO SPECIES OF CYCLAMEN

1.	Stamens exserted from the mouth of the corolla; leaf-lamina with even, broad, triangular lobes	**C. rohlfsianum**
1.	Stamens included within the corolla; leaf-lamina not as above	**2**
2.	Corolla auriculate towards the base where the corolla-lobes reflex; flowering in the summer, autumn and early winter	**3**
2.	Corolla not auriculate; flowering mostly in the late winter, spring and summer	**8**
3.	Fruiting pedicels coiling in two directions from near the centre, or from near the base upwards; tuber with thick anchorage roots	**C. graecum**
3.	Fruiting pedicels coiling from the top downwards; tuber without thick anchorage roots	**4**
4.	Leaves orbicular or reniform, neither lobed nor angled along the margin; corolla plain, unmarked	**5**
4.	Leaves usually cordate, lobed and/or angled along the margin; corolla with basal markings to each lobe (if pure white then the leaf characters apply)	**6**
5.	Leaf-lamina very thick and fleshy with a distinct but finely toothed, rather beaded, margin and diverging basal lobes; corolla 11–15 mm long	**C. colchicum**
5.	Leaf-lamina thin and generally with an indistinctly toothed, not beaded, margin and converging or overlapping basal lobes; corolla 17–25 mm long	**C. purpurascens**
6.	Corolla with an M-shaped blotch towards the base of each lobe; calyx-lobes narrow-triangular, acuminate; tuber rooting from one side of the base	**C. cyprium**
6.	Corolla with a V-shaped blotch at the base of each lobe (not present in albinos); calyx-lobes broad-triangular, with an abrupt cuspidate apex; tuber rooting primarily from the top and sides	**7**
7.	Petioles and pedicels straight, arising straight above the tuber; corolla-lobes 18–35 mm long; tuber generally concave above	**C. africanum**
7.	Petioles and pedicels with a distinct elbow in the lower half, arising to the side of the tuber; corolla-lobes 14–22 mm long; tuber flat or somewhat convex above	**C. hederifolium**
8.	Pedicels thickening and curving downwards in fruit but not coiling	**9**
8.	Pedicels coiling from the top downwards as the fruit develop	**10**

84

9. Corolla-lobes 10–15 mm long; margin of leaf-lamina somewhat angled in the lower half · **C. somalense**

9. Corolla-lobes 20–37 mm long; margin of leaf-lamina not angled · **C. persicum**

10. Calyx-lobes 1-veined; corolla plain, unmarked or with a coloured zone around the nose; anthers never aristate · · · · · · · · · · **11**

10. Calyx-lobes 3–5-veined; corolla with a dark blotch or mark at the base of each lobe or, if unmarked, then corolla glandular in part and anthers aristate · **15**

11. Leaf-lamina thin, cordate, generally lobed and/or angled; tuber smooth and velvety · **12**

11. Leaf-lamina thicker and more fleshy, unlobed and unangled; tuber corky · **14**

12. Corolla pink to deep magenta overall, or white or pale pink with a deeper pink zone around the nose; leaf-lamina with a deep bright green base colour · **C. repandum**

12. Corolla plain, white or very pale pink; leaf-lamina generally with a grey-green base colour · **13**

13. Leaf-lamina with an acute apex, the margin flat; corolla-lobes 15–26 mm long · **C. creticum**

13. Leaf-lamina with an obtuse apex, the margin usually somewhat revolute; corolla-lobes 9–16 mm long · · · · · · · · · · · · · **C. balearicum**

14. Leaf-lamina very thick and fleshy with a distinct but finely toothed, rather beaded, margin and diverging basal lobes; corolla 11–15 mm long · **C. colchicum**

14. Leaf-lamina thin and generally with an indistinctly toothed, not beaded, margin and converging or overlapping basal lobes; corolla 17–25 mm long · **C. purpurascens**

15. Flowers appearing in the late summer and autumn; anthers always aristate · **16**

15. Flowers appearing in the winter or spring; anthers aristate or not · **18**

16. Corolla-lobes distinctly toothed towards the apex; leaf-lamina often flushed with pink or red above when young · · · · · · · · **C. mirabile**

16. Corolla-lobes indistinctly toothed to entire; leaf-lamina rarely flushed with pink or red above · · · · · · · · · · · · · · · · · **17**

17. Leaf-lamina longer than wide, usually toothed; corolla-lobes with a distinctive dark blotch towards the base, 14–19 mm long (if pure white then leaf characters apply) · · · · · · · · · · · · · · · **C. cilicium**

17. Leaf-lamina as wide as long or wider, often entire; corolla lobes plain, 10–16 mm long · · · · · · · · · · · · · · · · · · · **C. intaminatum**

18. Anthers not aristate; tuber corky, rooting from all over the base: corolla glandular, the lobes entire 19
18. Anthers aristate; tuber velvety, rooting only from the centre of the base; corolla rarely glandular, the lobes usually somewhat toothed 20

19. Mouth of corolla 10–13 mm diameter; corolla pink to deep magenta; leaf-lamina angled but not toothed **C. libanoticum**
19. Mouth of corolla 3–6 mm diameter, corolla very pale rose pink or whitish; leaf-lamina toothed but not angled **C. pseudibericum**

20. Blotch at base of the corolla-lobes with a pair of paler 'eyes'; flowers generally unscented **C. coum**
20. Blotch at base of corolla-lobes solid or consisting of close more or less parallel lines; flowers sweetly scented 21

21. Leaf-lamina marbled or variously patterned above, somewhat toothed; corolla-lobes 9–13 mm long, always more or less horizontal, twisting through only 90°, the blotch solid **C. trochopteranthum**
21. Leaf-lamina plain, more or less entire; corolla-lobes 4–11 mm long, often erect, twisting through 180°, the blotch consisting of close lines **C. parviflorum**

KEY TO SPECIES OF GALANTHUS

1.	Leaves bright to dark green, or very slightly mat	**2**
1.	Leaves glaucous or glaucescent	**9**
2.	Inner perianth segments marginate (without an apical notch)	**3**
2.	Inner perianth segments emarginate (with an apical notch)	**4**
3.	Anthers blunt; inner perianth segments obtuse at the apex	**G. platyphyllus**
3.	Anthers apiculate; inner perianth segments acute at the apex	**G. krasnovii**
4.	Leaves applanate in vernation	**5**
4.	Leaves supervolute in vernation	**6**
5.	Leaves usually bright, shining, green or infrequently mat, lacking a median stripe	**G. lagodechianus**
5.	Leaves usually mat, often with a faint median stripe	**G. rizehensis**
6.	Inner perianth segments with 2 green marks (1 apical and 1 basal)	**G. fosteri**
6.	Inner perianth segments with 1 green mark (apical only)	**7**
7.	Inner perianth mark up to, but never greater than, half the length of the segment; leaves green or very slightly mat; Caucasus and NE Turkey	**8**
7.	Inner perianth mark at least half the length of the segment, and usually greater; leaves rather mat; the Aegean Islands of Greece	**G. ikariae**
8.	Inner perianth mark ± ∧ to ∩-shaped, rounded at the apex; leaves matt green; E Caucasus and N Iran	**G. transcaucasicus**
8.	Inner perianth mark ± ∩-shaped, usually flat-topped at the apex; leaves bright green or glossy green; W and central Caucasus and NE Turkey	**G. woronowii**
9.	Leaves either glaucescent, or leaf surfaces discolorous	**10**
9.	Leaves distinctly glaucous, leaf surfaces concolorous	**13**
10.	Leaves explicative (2-folded) in vernation; leaf-lamina folded towards the abaxial surface of the leaf	**G. plicatus**
10.	Leaves applanate in vernation; leaf-lamina not folded	**11**
11.	Upper leaf surface with a prominent glaucous median stripe, on a green or glaucescent background, abaxial and adaxial leaf surfaces discolorous	**G. reginae-olgae**
11.	Upper leaf surface glaucescent, without a distinct glaucous median stripe, abaxial and adaxial leaf surfaces concolorous	**12**

12. Leaves absent or distinctly shorter than scape at flowering time; autumn flowering — **G. peshmenii**
12. Leaves well developed at flowering time, slightly shorter to longer than the scape at flowering; winter to spring flowering — **G. nivalis**

13. Inner perianth segments with 2 separate green marks, 1 apical and the other basal, or 1 large ± X-shaped mark — **14**
13. Inner perianth segments with 1 green mark at the apex, or rarely a very faint yellow or green coloration near the base — **15**

14. Leaves supervolute in vernation, 1–3.5 cm wide — **G. elwesii**
14. Leaves applanate in vernation, 0.3–0.8(–1) cm wide — **G. gracilis**

15. Abaxial leaf surface with distinct longitudinal furrows — **G. koenenianus**
15. Abaxial leaf surface ± smooth, without longitudinal furrows — **16**

16. Leaves supervolute in vernation, usually more than 1 cm wide — **18**
16. Leaves applanate in vernation, less than 1 cm wide — **17**

17. Leaves usually more than 0.6 cm wide; autumn to winter flowering; S Turkey — **G. cilicicus**
17. Leaves less than 0.5 cm wide; spring flowering; N Caucasus — **G. angustifolius**

18. Inner perianth segments with apical markings only, i.e. 1 mark on each inner perianth segment, basal markings absent (i.e. always only 1 mark on each inner perianth segment); leaves usually narrow c. 1–2 cm wide; Caucasus, Transcaucasus and NE Turkey — **G. alpinus**
18. Inner perianth segments with apical and basal markings, i.e. 2 marks on each inner perianth segment, basal markings sometimes absent (i.e. with 1 mark on each inner perianth segment); leaves usually broad c. 2–3.5 cm wide; SE Europe, the Balkans and Turkey — **G. elwesii**

KEY TO SUBSPECIES AND VARIETIES OF GALANTHUS

Key to the subspecies of Galanthus plicatus

1. Inner perianth segments with 1 mark at the apex   subsp. **plicatus**
1. Inner perianth segments with 2 marks, 1 apical and 1 basal   subsp. **byzantinus**

Key to the subspecies of Galanthus reginae-olgae

1. Leaves absent, or distinctly shorter than scape during flowering;
    autumn flowering (September to December)   subsp. **reginae-olgae**
1. Leaves never absent, always well developed during flowering;
    winter to spring flowering (January to March)   subsp. **vernalis**

Key to varieties of Galanthus alpinus

1. Bulb-scales whitish; flowers either ellipsoid or globose; seed
    capsules developing to maturity (fertile); widespread in the
    Caucasus (Russia, Georgia, Armenia) and neighbouring areas (NE
    Turkey)   var. **alpinus**
1. Bulb-scales yellowish; flowers globose; mature seed capsules not
    known to develop to maturity (sterile); only found in one site in
    district Chegem, upper Kamenka region, Karbardino-Balcaria
    (Russia)   var. **bortkewitschianus**

KEY TO SPECIES OF STERNBERGIA

1.	Flowers produced in spring	2
1.	Flowers produced in autumn	3
2.	Flowers white	**S. candida**
2.	Flowers yellow	**S. fischeriana**
3.	Perianth tube usually 2–6.5 cm long; leaves absent at flowering time	4
3.	Perianth tube 2 cm or less long; leaves appearing at or before flowering time	5
4.	Perianth segments 3.5–7.5 cm long; leaves 8–16 mm wide, grey-green	**S. clusiana**
4.	Perianth segments 3 cm or less long; leaves 1–4 mm wide, dark green	**S. colchiciflora**
5.	Leaves bright, shiny green, flat in cross section	6
5.	Leaves deep green with a greyish median stripe, channelled in cross-section	7
6.	Leaves 7–12 mm wide; perianth segments 3–3.5 cm long	**S. lutea**
6.	Leaves 2–5 mm wide; perianth segments 2–3 cm long	**S. greuteriana**
7.	Perianth tube 0.4–1 cm long	**S. sicula**
7.	Perianth tube 1.5–2 cm long	8
8.	Perianth segments 3.5–4 cm long	**S. schubertii**
8.	Perianth segments 1.5–1.8 cm long	**S. pulchella**

Explanation of terms used in keys

abaxial the side or face positioned away from the axis; dorsal or lower surface
acuminate having a gradually diminishing point
acute sharp, ending in a point
adaxial the side or face next to the axis; ventral or upper surface
apex the tip or point
apiculum a sharp and short, but not stiff point
apiculate furnished with an apiculum
applanate (vernation) with both leaves flat together in bud; the adaxial surfaces facing each other in bud
aristate with a hair-like appendage
auricle an ear-shaped lobe or appendage, for example: at the base of the corolla or leaf
auriculate possessing auricles
calyx the outer whorl of the floral parts, i.e. below the perianth or corolla; usually green and not brightly coloured

cordate heart-shaped

concolorous of the same colour, uniform in colour

corolla inner whorl of the floral parts, composed of petals; usually brightly coloured

cuspidate with an apex sharply constricted into an elongated, sharp-pointed tip

discolorous not the same colour; not uniform in colour

ellipsoid an elliptic solid: oblong with regularly rounded ends

emarginate margin not entire (complete), e.g. with a notch

entire with a continuous margin; not indented or toothed, whole

exserted protruding; stamens or style exposed, i.e. longer than the petals or inner perianth segments

explicative (vernation) the leaves folded sharply backwards in bud, so that the surfaces of the folded parts are brought together

fruiting pedicel the fruit stalk

glandular having secreting organs or glands

glaucescent almost (becoming) glaucous; greyish-green

glaucous covered with bluish or grey bloom, like the wax covering on a plum

globose nearly spherical (round)

leaf-lamina the leaf blade

marginate with a complete margin; without notches

matt lacking lustre, dull

median in the middle, running lengthways

obtuse blunt, rounded

pedicel the stalk of the flower or fruit

perianth the showy parts of the flower: the petals (corolla) or calyx, or both together; in *Galanthus* and *Sternbergia* (and many other monocotyledons) there is no calyx, and the flower is referred to in terms of its perianth segments instead of petals

perianth tube the part of the perianth where it is fused to form a tube, usually at the base

perianth segment one part of the perianth; in *Galanthus* and *Sternbergia* there are six segments to each perianth (the flower)

petiole the leaf stalk

reniform kidney-shaped

revolute rolled back from the margin or apex

reflexed abruptly bend downward or backwards

scape a leafless stalk arising from the ground, bearing one or many flowers

supervolute (vernation) with one leaf enveloping the other in bud, or partially so: the leaves rolled towards the adaxial (upper) surface in bud

tuber a swollen storage organ, subterranean or partly so

vernation the order of unfolding of leaf buds; in *Galanthus* it is possible to ascertain the type of vernation after the bud stage and during maturity, by looking at the how the leaves are folded at the base of the plant

CLÉS DE DÉTERMINATION POUR CYCLAMEN, GALANTHUS ET STERNBERGIA

CLE POUR LES ESPECES DE CYCLAMEN

1.	Etamines exsertes; limbe à lobes égaux, larges et triangulaires	**C. rohlfsianum**
1.	Etamines non exsertes; limbe différent de ci-dessus	**2**
2.	Corolle auriculée vers la base, où les lobes de la corolle se replient; floraison en été, en automne et au début de l'hiver	**3**
2.	Corolle non auriculée, floraison principalement à la fin de l'hiver, au printemps et en été	**8**
3.	Pédicelles fructifères s'enroulant dans deux directions à partir du centre, ou de la base vers le haut; tubercule présentant d'épaisses racines d'ancrage	**C. graecum**
3.	Pédicelles fructifères s'enroulant du haut vers le bas; tubercule ne présentant pas d'épaisses racines d'ancrage	**4**
4.	Feuilles orbiculaires ou réniformes, jamais lobées ni découpées le long de la marge; corolle unicolore sans marques	**5**
4.	Feuilles généralement cordiformes, lobées et/ou découpées le long de la marge; corolle présentant des marques à la base de chaque lobe (si blanc pur, voir les caractères de la feuille)	**6**
5.	Limbe très épais et charnu, à marge nettement et finement dentelée et ourlée et aux lobes de la base divergents; corolle de 11–15 mm de long	**C. colchicum**
5.	Limbe mince généralement à marge à dents peu visibles et sans ourlet, aux lobes de la base convergents ou recouvrants; corolle de 17–25 mm de long	**C. purpurascens**
6.	Corolle présentant une tache en forme de "M" vers la base de chaque lobe; lobes du calice étroitement triangulaires et acuminés; tubercule émettant des racines d'un seul côté de la base	**C. cyprium**
6.	Corolle présentant une tache en forme de "V" vers la base de chaque lobe (absent chez les albinos); lobes du calice largement triangulaires avec une extrémité brusquement cuspide; tubercule émettant des racines principalement depuis le sommet et les côtés	**7**
7.	Pétioles et pédicelles droits, poussant directement au-dessus du tubercule; lobes de la corolle de 18–35 mm de long; tubercule généralement concave sur le dessus	**C. africanum**
7.	Pétioles et pédicelles formant un coude bien visible sur la moitié inférieure, poussant depuis le côté du tubercule; lobes de la corolle de 14–22 mm de long; tubercule plat et un peu convexe sur le dessus	**C. hederifolium**

8. Pédicelles s'élargissant et se repliant vers le bas dans le fruit mais sans s'enrouler **9**
8. Pédicelles s'enroulant à partir du sommet vers la base à la maturation du fruit **10**

9. Lobes de la corolle de 10–15 mm de long; marge de la feuille formant un angle dans sa partie inférieure **C. somalense**
9. Lobes de la corolle de 20–37 mm de long; marge de la feuille ne formant pas d'angle **C. persicum**

10. Lobes du calice uninervés, corolle unicolore sans marques ou présentant une zone colorée autour de l'appendice; anthères jamais arristées **11**
10. Lobes du calice à 3–5 nervures, corolle présentant une tache sombre ou une marque à la base de chaque lobe ou, s'il n'y a pas de marques, corolle glanduleuse par endroit et anthères arristés **15**

11. Limbe mince, cordiforme et généralement lobé et/ou faisant un angle; tubercule lisse et velouté **12**
11. Limbe plus épais et charnu, sans lobes ni angles; tubercule subéreux **14**

12. Corolle entièrement rose à magenta foncé, ou blanche ou rose pâle avec des zones roses plus sombres autour de l'appendice; limbe à couleur de fond vert brillant **C. repandum**
12. Corolle unicolore, blanche ou rose très pâle; limbe généralement gris-vert **13**

13. Limbe à extrémité pointue, marge plane; lobes de la corolle de 15–26 mm de long **C. creticum**
13. Limbe à extrémité arrondie, marge généralement un peu retournée; lobes de la corolle de 9–16 mm de long **C. balearicum**

14. Limbe très épais et charnu à marge finement mais nettement dentelée, avec un ourlet et lobes de la base divergents; corolle de 11–15 mm de long **C. colchicum**
14. Limbe mince ayant généralement une marge à dents peu visibles et sans ourlet, lobes de la base convergents ou recouvrants; corolle de 17–25 mm de long **C. purpurascens**

15. Floraison à la fin de l'été et en automne; anthères toujours aristées **16**
15. Floraison en hiver ou au printemps, anthères aristées ou non **18**

16. Lobes de la corolle distinctement dentés vers l'extrémité, jeune limbe souvent teinté de rose ou de rouge **C. mirabile**
16. Lobes de la corolle indistinctement dentés à entiers, limbe rarement teinté de rose ou de rouge **17**

94

17. Limbe plus long que large, généralement denté; lobes de la corolle présentant une marque nette foncée vers la base, de 14–19 mm de long (si fl. blanches, voir caractères de la feuille) **C. cilicium**
17. Limbe aussi long que large ou plus large, souvent entier; lobes de la corolle unicolores, de 10–16 mm de long **C. intaminatum**

18. Anthères non aristées; tubercule subéreux, émettant des racines tout autour de la base; corolle glanduleuse; lobes entiers **19**
18. Anthères aristées; tubercule velouté, émettant des racines seulement à partir du centre de la base; corolle rarement glanduleuse; lobes généralement un peu dentelés **20**

19. Corolle de 10–13 mm de diamètre, rose à magenta foncé; limbe découpé mais non denté **C. libanoticum**
19. Corolle de 3–6 mm de diamètre, rose très pâle à blanchâtre; limbe denté mais non découpé **C. pseudibericum**

20. Tache à la base des lobes de la corolle avec une paire d'"yeux" plus pâles; fleurs généralement inodores **C. coum**
20. Tache à la base des lobes de la corolle unie ou formée de lignes rapprochées plus ou moins parallèles; fleurs à odeur douce **21**

21. Limbe tacheté ou avec des motifs variables sur la face supérieure, un peu denté; lobes de la corolle de 9–13 mm de long, toujours plus ou moins horizontaux, se recourbant seulement de 90°; tache unie **C. trochopteranthum**
21. Limbe unicolore, plus ou moins entier; lobes de la corolle de 4–11 mm de long, souvent dressés, se recourbant à 180°; tache formée de lignes rapprochées **C. parviflorum**

CLÉ POUR LES ESPÈCES DE GALANTHUS

1. Feuilles vert vif à vert foncé, ou très légèrement mates   **2**
1. Feuilles glauques ou devenant glauques   **9**

2. Segments du périanthe intérieur marginés (sans entaille apicale)   **3**
2. Segments du périanthe intérieur émarginés (avec entaille apicale)   **4**

3. Anthères obtus; segments du périanthe intérieur obtus à l'extrémité   **G. platyphyllus**
3. Anthères apiculés, segments du périanthe intérieur aigus à l'extrémité   **G. krasnovii**

4. Feuilles aplaties en vernation   **5**
4. Feuilles enroulées en vernation   **6**

5. Feuilles généralement vert vif et luisantes ou rarement mates, sans bande médiane   **G. lagodechianus**
5. Feuilles généralement mates, présentant souvent une bande médiane délavée   **G. rizehensis**

6. Segments du périanthe intérieur présentant 2 marques vertes (1 apicale et 1 basale)   **G. fosteri**
6. Segments du périanthe intérieur présentant 1 marque verte apicale   **7**

7. Marque du périanthe intérieur pouvant atteindre (sans jamais dépasser) la moitié de la longueur du segment; feuilles vertes ou très légèrement mates; Caucase et NE de la Turquie   **8**
7. Marque du périanthe intérieur atteignant au moins la moitié de la longueur du segment, et généralement plus grande; feuilles plutôt mates; îles de la mer Egée (Grèce)   **G. ikariae**

8. Marque du périanthe intérieur ± en forme de ∧ ou de ∩, arrondi à l'extrémité; feuilles vert mat; E du Caucase et N de l'Iran   **G. transcaucasicus**
8. Marque du périanthe intérieur ± en forme de ∩, généralement à bord aplati; feuilles vert vif ou brillantes; O et centre du Caucase et NE de la Turquie   **G. woronowii**

9. Feuilles plutôt glauques ou discolores   **10**
9. Feuilles nettement glauques, concolores   **13**

10. Feuilles repliées (en deux) en vernation; limbe replié vers la surface abaxiale de la feuille   **G. plicatus**
10. Feuilles aplatie en vernation; limbe non replié   **11**

11. Face supérieure de la feuille présentant une ligne médiane glauque nette sur fond vert plutôt glauque; faces abaxiales et adaxiales discolores — **G. reginae-olgae**

11. Face supérieure de la feuille glauque, sans ligne médiane nette, surfaces abaxiales et adaxiales concolores — **12**

12. Pas de feuilles ou feuilles nettement plus courtes que la hampe florale à la floraison; floraison en automne — **G. peshmenii**

12. Feuilles bien développées à la floraison; floraison en hiver et au printemps — **G. nivalis**

13. Segments du périanthe intérieur présentant 2 marques vertes distinctes, 1 apicale et 1 basale, ou 1 grande marque plus ou moins en forme de X — **14**

13. Segments du périanthe intérieur présentant 1 marque verte à l'extrémité, ou rarement, une très légère coloration jaune ou verte à la base — **15**

14. Feuilles enroulées en vernation, de 1–3,5 cm de large — **G. elwesii**

14. Feuilles aplaties en vernation, de 0,3–0,8 cm de large — **G. gracilis**

15. Surface abaxiale de la feuille présentant des sillons longitudinaux bien distincts — **G. koenenianus**

15. Surface abaxiale de la feuille plus ou moins lisse, sans sillons longitudinaux — **16**

16. Feuilles enroulées en vernation, De 1–2 cm ou de plus de 2 cm de large — **18**

16. Feuilles aplaties en vernation, de moins d'1 cm de large — **17**

17. Feuilles généralement de 0,6 cm de large; floraison en automne et hiver; S de la Turquie — **G. cilicicus**

17. Feuilles de moins de 0,5 cm de large; floraison au printemps; N du Caucase — **G. angustifolius**

18. Segments du périanthe intérieur présentant des marques uniquement apicales, soit 1 seule marque sur chaque segment du périanthe intérieur; absence de marques basales; feuilles généralement d'1–2 cm de large; Caucase, région transcaucasienne et NE de la Turquie — **G. alpinus**

18. Segments du périanthe intérieur présentant des marques apicales et basales, soit 2 marques sur chaque segment du périanthe intérieur; parfois absence de la marque basale (dans ce cas, 1 seule marque sur chaque segment du périanthe intérieur); feuilles généralement de 2–3,5 cm de large; SE de l'Europe, Balkans et Turquie — **G. elwesii**

CLÉ POUR LES SOUS-ESPÈCES ET VARIÉTÉS DE GALANTHUS

Clé pour les sous-espèces de Galanthus plicatus

1.	Segments du périanthe intérieur présentant 1 marque à l'extrémité	subsp. **plicatus**
1.	Segments du périanthe intérieur présentant 2 marques: 1 basale, 1 apicale	subsp. **byzantinus**

Clé pour les sous-espèces de Galanthus reginae-olgae

1.	Feuilles absentes ou notablement plus courtes que la hampe florale lors de la floraison; floraison en automne (septembre à décembre)	subsp. **reginae-olgae**
1.	Feuilles jamais absentes, toujours bien développées lors de la floraison; floraison en hiver/printemps (janvier à mars)	subsp. **vernalis**

Clé pour les variétés de Galanthus alpinus

1.	Ecailles du bulbe blanchâtres, fleurs soit ellipsoïdes, soit globuleuses; graines se développant jusqu'à maturité (fertiles); répandu dans le Caucase (Russie, Géorgie, Arménie) et régions avoisinantes (NE de la Turquie)	var. **alpinus**
1.	Ecailles du bulbe jaunâtres, fleurs globuleuses; capsules de graines ne semblant pas se développer jusqu'à maturité (stérile); une seule station connue dans le district de Chegem, dans la région du Haut-Kamenka, Karbardino-Balcaria (Russie)	var. **bortkewitschianus**

CLÉ POUR LES ESPÈCES DE STERNBERGIA

1. Floraison au printemps   **2**
1. Floraison en automne   **3**

2. Fleurs blanches   **S. candida**
2. Fleurs jaunes   **S. fischeriana**

3. Tube du périanthe généralement de 2–6,5 cm de long; feuilles
    absentes lors de la floraison   **4**
3. Tube du périanthe généralement de moins de 2 cm de long; feuilles
    apparaissant avant ou pendant la floraison   **5**

4. Segments du périanthe de 3,5–7,5 cm de long; feuilles de 8–16 mm
    de large, gris-vert   **S. clusiana**
4. Segments du périanthe pouvant atteindre 3 cm de long; feuilles de 1–
    4 mm de large, vert foncé   **S. colchiciflora**

5. Feuilles brillantes, vert vif, planes en section transversale   **6**
5. Feuilles vert foncé présentant une ligne médiane grisâtre, cannelée en
    section transversale   **7**

6. Feuilles de 7–12 mm de large; segments du périanthe de 3–3,5 cm de
    long   **S. lutea**
6. Feuilles de 2–5 mm de large; segments du périanthe de 2–3 cm de
    long   **S. greuteriana**

7. Tube du périanthe de 0,4–1 cm de long   **S. sicula**
7. Tube du périanthe de 1,5–2 cm de long   **8**

8. Segments du périanthe de 3,5–4 cm de long   **S. schubertii**
8. Segments du périanthe de 1,5–1,8 cm de long   **S. pulchella**

Explication des termes utilisés dans les clés

abaxiale: (face) opposée à l'axe; dans la feuille, c'est la face inférieure
acuminé: en pointe
adaxiale: (face) du côté de l'axe; dans la feuille, c'est la face supérieure
aigu: pointu, se terminant en pointe
apiculé: terminé par un apiculum
apiculum: pointe aiguë et courte mais non rigide
aplati (en vernation): aux deux feuilles à plat l'une contre l'autre dans le bourgeon, faces
adaxiales se faisant face
apex: extrémité ou pointe
aristé: muni d'une arête
auricule: lobe ou appendice en forme d'oreille, par ex.; à la base de la corolle ou de la feuille

Annexe II: Cles de determination - *Sternbergia*

auriculé: présentant des auricules

calice: verticille extérieur des pièces florales, placé sous le périanthe ou la corolle, généralement vert et de couleur peu voyante

concolore: de la même couleur, de couleur uniforme

cordiforme: en forme de cœur

corolle: verticille intérieur des pièces florales, composé des pétales; généralement de couleur vive

cuspide: présentant une extrémité fortement rétrécie en une pointe longue et aiguë

discolore de couleurs différentes; de couleur non uniforme

ellipsoïde: de forme elliptique, oblong, aux bords arrondis régulièrement

émarginé: à marge non entière (incomplète), avec une entaille, par exemple

enroulée (en vernation): qualifie une feuille enveloppant plus ou moins entièrement une autre feuille dans le bourgeon; les feuilles sont enroulées vers la face adaxiale (supérieure) dans le bourgeon

entier: à marge continue, ni découpée, ni dentée; d'une pièce

exserts: qualifie les étamines ou le style plus longs que les pétales ou que les segments internes du périanthe

glanduleux: présentant des organes de sécrétion ou des glandes

glauque: couvert d'un duvet bleuté ou gris-vert, comme la pruine de la prune

globuleux: presque sphérique

hampe florale: tige sans feuille sortant du sol, portant une ou plusieurs fleurs

limbe: partie plane, élargie, de la feuille

marginé: à marge entière, sans entailles

médian: au milieu, dans le sens de la longueur

obtus: arrondi, non aigu

pédicelle fructifère: hampe du fruit

pédicelle: tige de la fleur ou du fruit

périanthe: pièces voyantes de la fleur: pétales (corolle) ou calice, ou les deux ensemble; chez *Galanthus* et *Sternbergia* (et beaucoup d'autres monocotylédones) il n'y a pas de calice et l'on parle de segments du périanthe plutôt que de pétales

pétiole: tige de la feuille

réniforme: en forme de rein

repliées (en vernation): qualifie les feuilles fortement repliées en arrière dans le bourgeon, les parties repliées se rejoignant

retourné: enroulé en arrière à partir de la marge ou de l'extrémité

rétracté: abruptement tordu vers le bas ou en arrière

segment du périanthe: partie du périanthe; chez *Galanthus* et *Sternbergia*, il y a six segments pour chaque périanthe (la fleur)

subéreux: à consistance de bouchon

tube du périanthe: partie du périanthe soudée, généralement à la base, pour former un tube

tubercule: organe de stockage renflé, partiellement ou entièrement souterrain

vernation (ou préfoliaison): ordre dans lequel les feuilles du bourgeon se déplient; chez *Galanthus*, on reconnaît le type de vernation après le débourrage et à la maturité des feuilles en observant la manière dont les feuilles sont insérées à la base de la plante

CLAVES PARA LA IDENTIFICACION DE CYCLAMEN, GALANTHUS Y STERNBERGIA

CLAVES PARA LAS ESPECIES DE CYCLAMEN

1.	Estambres exertos de la boca de la corola; lámina foliar con lóbulos uniformes, anchos y triangulares	**C. rohlfsianum**
1.	Estambres incluidos en la corola; lámina foliar distinta de la anterior	**2**
2.	Corola auriculada hacia la base, donde los lóbulos de la corola son reflexos; floración en verano, otoño y principios del invierno	**3**
2.	Corola no auriculada; floración por lo general a finales del invierno, en primavera y en verano	**8**
3.	Pedicelos fructíferos ascienden enrollados en dos sentidos desde la región central o basal; tubérculos con raíces gruesas de fijación	**C. graecum**
3.	Pedicelos fructíferos descienden enrollados; el tubérculo no tiene raíces gruesas de fijación	**4**
4.	Hojas orbiculares o reniformes, sin lóbulos ni ángulos en el margen; corola lisa, sin marcas	**5**
4.	Hojas normalmente cordatas, con lóbulos y/o ángulos en el margen; corola con marcas basales en cada lóbulo (si son blanco puro, los caracteres de la hoja se aplican)	**6**
5.	Lámina foliar muy gruesa y carnosa con un margen distinto pero finamente dentado, casi festoneado, y lóbulos basales divergentes; corola 11–15 mm de largo	**C. colchicum**
5.	Lámina foliar delgada, por lo general con un margen indistinto dentado, no festoneado, que converge con los lóbulos basales o se superpone a ellos; corola 17–25 mm de largo	**C. purpurascens**
6.	Corola con una mancha en forma de M cerca de la base de cada lóbulo; lóbulos del cáliz en forma de triángulo de base angosta, acuminados; las raíces en tubérculo parten de un lado de la base	**C. cyprium**
6.	Corola con una mancha en forma de V en la base de cada lóbulo (excepto en los albinos); lóbulos del cáliz en forma de triángulo de base ancha con ápice cuspidado abrupto; raíces del tubérculo parten primordialmente de la parte superior y los lados	**7**
7.	Pecíolos y pedicelos derechos, parten directamente del tubérculo; lóbulos de la corola 18–35 mm de largo; tubérculo generalmente cóncavo en la parte superior	**C. africanum**
7.	Pecíolos y pedicelos con un codo bien marcado en la mitad inferior que parte del lado del tubérculo, lóbulos de la corola 14–22 mm de largo; tubérculo plano o ligeramente convexo en la parte superior	**C. hederifolium**

8. Los pedicelos se engrosan y curvan hacia abajo en el fruto pero sin enrollarse — **9**
8. Los pedicelos descienden enrollados a medida que se desarrolla el fruto — **10**

9. Lóbulos de la corola 10–15 mm de largo; margen de la lámina foliar ligeramente angulosa en la mitad inferior — **C. somalense**
9. Lóbulos de la corola 20 37 mm de largo; margen de la lámina foliar sin Ángulos — **C. persicum**

10. Lóbulos del cáliz con 1 nervadura; corola lisa, sin marcas o con una zona de color alrededor de la nariz; anteras nunca aristadas — **11**
10. Lóbulos del cáliz con 3–5 nervaduras; corola con una marca o señal oscura en la base de cada lóbulo o, si no hay marcas, corola glandular en parte y anteras aristadas — **15**

11. Lámina foliar delgada, cordata, generalmente lobulada y/o angulosa; tubérculo liso y aterciopelado — **12**
11. Lámina foliar más gruesa y carnosa, sin lóbulos ni ángulos; tubérculos uberosos — **14**

12. Corola color rosado a violáceo oscuro integramente, o blanco o rosa pálido con una zona rosa más oscuro alrededor de la nariz; lámina foliar con una base de color verde oscuro vivo — **C. repandum**
12. Corola lisa, blanca o rosa muy pálido; lámina foliar generalmente con una base de color verde grisáceo — **13**

13. Lámina foliar con ápice agudo y margen llano; lóbulos de la corola 15–26 mm de largo — **C. creticum**
13. Lámina foliar con ápice obtuso, margen por lo general ligeramente revoluto; lóbulos de la corola 9–16 mm de largo — **C. balearicum**

14. Lámina foliar muy gruesa y carnosa con un margen distinto pero finamente dentado, casi festoneado y lóbulos basales divergentes; corola 11–15 mm largo — **C. colchicum**
14. Lámina foliar delgada, por lo general con un margen indistinto dentado, no festoneado, y con lóbulos basales convergentes o superpuestos; corola 17–25 mm de largo — **C. purpurascens**

15. Floración al final del verano y en otoño; anteras siempre aristadas — **16**
15. Floración en invierno o primavera; anteras aristadas o no — **18**

16. Lóbulos de la corola nítidamente dentados hacia el ápice; la lámina foliar joven suele ser rosa o rojo en la parte superior — **C. mirabile**
16. Lóbulos de la corola indistintamente dentados o enteros; lámina foliar rara vez es rosa o rojo en la parte superior — **17**

17. Lámina foliar más larga que ancha, por lo general dentada; lóbulos de la corola con una mancha oscura distintiva en la zona basal, 14–19 mm de largo (si son blanco puro, los caracteres de la hoja se aplican) **C. cilicium**

17. Lámina foliar tan larga como ancha o más ancha, por lo general entera; Lóbulos de la corola lisos, 10–16 mm de largo **C. intaminatum**

18. Anteras no aristadas; tubérculo suberoso, las raíces parten de toda la zona basal; corola glandular con lóbulos enteros **19**

18. Anteras aristadas; tubérculo aterciopelado, las raíces parten del centro de la la zona basal; corola rara vez glandular, los lóbulos suelen ser ligeramente dentados **20**

19. Boca de la corola 10–13 mm de diámetro; corola de color rosado a violáceo oscuro; lámina foliar angulosa pero no dentada **C. libanoticum**

19. Boca de la corola 3–6 mm de diámetro; corola de un rosa muy pálido o blanquecina; lámina foliar dentada pero no angulosa **C. pseudibericum**

20. Mancha en la base de los lóbulos de la corola con un par de 'ojos' más claros; flores por lo general sin aroma **C. coum**

20. Mancha en la base de los lóbulos de la corola lisa o formada por líneas juntas y más o menos paralelas; flores con aroma dulce **21**

21. Lámina foliar veteada o con distintos dibujos en la parte superior, ligeramente dentada; lóbulos de la corola 9–13 mm de largo, siempre más o menos horizontales, se tuercen sólo hasta 90º, la mancha es lisa **C. trochopteranthum**

21. Lámina foliar lisa, más o menos entera; lóbulos de la corola 4–11 mm de largo, con frecuencia empinados, se tuercen hasta 180º, la mancha está formada por líneas juntas **C. parviflorum**

CLAVE PARA LAS ESPECIES DE GALANTHUS

1.	Hojas verde vivo a verde oscuro, o muy ligeramente mate	**2**
1.	Hojas glaucas o glaucescentes	**9**
2.	Segmentos internos del perianto marginados (sin escotadura apical)	**3**
2.	Segmentos internos del perianto emarginados (con una escotadura apical)	**4**
3.	Anteras romas; segmentos internos del perianto obtusos en el ápice	**G. platyphyllus**
3.	Anteras apiculadas; segmentos internos del perianto agudos en el ápice	**G. krasnovii**
4.	Hojas aplanadas en vernación	**5**
4.	Hojas supervolutas en vernación	**6**
5.	Hojas por lo general verde brillante, lustroso, o rara vez mate, sin una lista media	**G. lagodechianus**
5.	Hojas por lo general mate, suelen tener una lista media tenue	**G. rizehensis**
6.	Los segmentos internos del perianto tienen 2 marcas verdes (1 apical y 1 basal)	**G. fosteri**
6.	Los segmentos internos del perianto tienen 1 marca verde (sólo apical)	**7**
7.	El perianto interno llega hasta la mitad de la longitud del segmento y nunca lo sobrepasa; hojas verdes o ligeramente mate; Cáucaso y noreste de Turquía	**8**
7.	El perianto interno llega por lo menos a la mitad de la longitud del segmento y por lo general es mayor; hojas más bien mate ; islas griegas en el mar Egeo	**G. ikariae**
8.	Marca en el perianto interno en forma ± de ∧ a ∩, redondeada en el ápice; hojas verde mate; este del Cáucaso y norte de Irán	**G. transcaucasicus**
8.	Marca en el perianto interno en forma ± de ∩, por lo general con el ápice plano; verde brillante o verde lustroso; oeste y centro del Cáucaso y noreste de Turquía	**G. woronowii**
9.	Hojas glaucescentes, o superficie de la hoja discolora	**10**
9.	Hojas claramente glaucas, o superficie de la hoja concolora	**13**
10.	Hojas explicativas (2 pliegues) en vernación; lámina foliar plegada hacia la superficie abaxial de la hoja	**G. plicatus**
10.	Hojas aplanadas en vernación; la lámina foliar no está plegada	**11**

11. Superficie ventral de la hoja con una lista media glauca prominente sobre un fondo glaucescente, superficies abaxial y adaxial de la hoja discoloras   **G. reginae-olgae**

11. Superficie ventral de la hoja glaucescente, sin lista media glauca distinta, superficie abaxial y adaxial de la hoja del mismo color   **12**

12. Hojas ausentes o claramente más cortas que el escapo en el momento de la floración; las flores aparecen en otoño   **G. peshmenii**

12. Hojas bien desarrolladas en el momento de la floración, apenas más cortas o más largas que el escapo en el momento de la floración; las flores aparecen en invierno y primavera   **G. nivalis**

13. Segmento del perianto interno con dos marcas verdes separadas, 1 apical y otra basal, o 1 marca grande en forma ± de X   **14**

13. Segmentos internos del perianto con 1 marca verde en el ápice, o rara vez una coloración amarillo tenue o verde cerca de la base   **15**

14. Hojas supervolutas en vernación, 1–3,5 cm ancho   **G. elwesii**

14. Hojas aplanadas en vernación, 0,3–0,8(–1) cm ancho   **G. gracilis**

15. Superficie abaxial de la hoja con surcos longitudinales distintos   **G. koenenianus**

15. Superficie abaxial de la hoja ± lisa, sin surcos longitudinales   **16**

16. Hojas supervolutas en vernación, hasta 2 cm o más de ancho   **18**

16. Hojas aplanadas en la vernación, menos de 1 cm de ancho   **17**

17. Hojas por lo general de más de 0,6 cm de ancho; floración de otoño a invierno; sur de Turquía   **G. cilicicus**

17. Hojas de menos de 0,5 cm de ancho; floración en primavera; norte del Cáucaso   **G. angustifolius**

18. Los segmentos internos del perianto tienen marcas apicales sólamente, es decir, hay 1 marca en cada segmento, no hay marcas basales (hay, pues, siempre 1 sola marca en cada segmento); las hojas suelen ser angostas, 1–2 cm de ancho; Cáucaso, Transcáucaso y noreste de Turquía   **G. alpinus**

18. Los segmentos internos del perianto tienen marcas apicales y basales, es decir, hay 2 marcas en cada segmento, a veces las marcas basales faltan (hay, pues, 1 marca en cada segmento); hojas en general anchas, 2–3,5 cm de ancho; sureste de Europe, los Balkanes y Turquía   **G. elwesii**

CLAVES PARA LAS SUBESPECIES Y VARIEDADES DE GALÁNTHUS

Clave para las subespecies de Galanthus plicatus

1.	Segmentos del perianto interno con 1 marca en el ápice	subsp. **plicatus**
1.	Segmentos del perianto interno con 2 marcas, 1 apical y 1 basal	subsp. **byzantinus**

Clave para las subespecies de Galanthus reginae-olgae

1.	Hojas ausentes, o mucho más cortas que el escapo durante la floración; las flores aparecen en otoño (septiembre a diciembre)	subsp. **reginae-olgae**
1.	Hojas nunca ausentes, siempre bien desarrolladas durante la floración; las flores aparecen en primavera (enero a marzo)	subsp. **vernalis**

Clave para las variedades de Galanthus alpinus

1.	Escamas del bulbo blanquecinas; flores elipsoides o globosas; las cápsulas de la semilla alcanzan la madurez fértil) abunda en el Cáucaso (Rusia, Georgia, Armenia) y zonas vecinas (noreste de Turquía)	var. **alpinus**
1.	Escamas del bulbo amarillentas; flores globosas; no se conocen cápsulas de la semilla que alcancen la madurez (estéril); sólo se encuentra en una localidad en el distrito de Chegem, región del alto Kamenka, Karbardino-Balcaria (Rusia)	var. **bortkewitschianus**

CLAVE PARA LAS ESPECIES DE STERNBERGIA

1.	Floración en primavera	**2**
1.	Floración en otoño	**3**
2.	Flores blancas	**S. candida**
2.	Flores amarillas	**S. fischeriana**
3.	Tubo del perianto normalmente 2–6,5 cm largo; hojas ausentes durante la floración	**4**
3.	Tubo del perianto 2 cm o menos de largo; las hojas aparecen durante la floración o antes	**5**
4.	Segmentos del perianto 3,5–7,5 cm largo; hojas 8–16 mm ancho, gris-verde	**S. clusiana**
4.	Segments del perianto 3 cm o menos de largo; hojas 1–4 mm de ancho, verde oscuro	**S. colchiciflora**
5.	Hojas verde brillante lustroso, llanas en corte transversal	**6**
5.	Hojas verde oscuro con una lista media grisácea, acanalada en corte transversal	**7**
6.	Hojas 7–12 mm ancho; segmentos del perianto 3–3,5 cm de largo	**S. lutea**
6.	Hojas 2–5 mm ancho; segmentos del perianto 2–3 cm largo	**S. greuteriana**
7.	Tubo del perianto 0,4–1 cm largo	**S. sicula**
7.	Tubo del perianto 1,5–2 cm largo	**8**
8.	Segmentos del perianto 3,5–4 cm largo	**S. schubertii**
8.	Segmentos del perianto 1,5–1,8 cm largo	**S. pulchella**

Explicación de los términos utilizados en las claves

abaxial: lado o porción más alejado del eje; superficie dorsal
acuminado: que se adelgaza gradualmente hasta formar una punta
agudo: punzante, termina en punta
adaxial: lado o porción próximo al eje; superficie ventral
apice: punta o terminación distal
apículo: punta aguda, corta y flexible
apiculado: terminado en un apículo
aplanado (vernación): ambas hojas pegadas en la yema; las superficies adaxiales frente a frente en la yema
aristado: con un apéndice en forma de cerda
aurícula: apéndice o lóbulo con forma de oreja, por ejemplo, en la base de la corola o la hoja

auriculado: que tiene aurículas

cáliz: el vertílico externo de las envolturas florales, es decir, debajo del perianto o la **corola:** normalmente verde mate

cordato: acorazonado

concoloro: del mismo color, unicoloro

corola: vertilicio interno de las envolturas florales, compuesto de pétalos, normalmente de colores brillantes

cuspidado: con un ápice marcadamente constricto hasta una punta aguda y alargada

discoloro: de dos o más colores, con colores diferentes

elipsoide: oval redondeado en los extremos

emarginado: con un margen incompleto, p.e.: con una muesca somera en el ápice

entero: con un margen continuo, no dentellado o dentado, completo

exerto: asomado; estambre o estilo expuestos, es decir, más largos que los pétalos o los segmentos internos del perianto

explicativo (vernación): hojas marcadamente dobladas hacia atrás en la yema, de manera que las superficies de las partes dobladas se tocan

pedicelo fructífero: el pedículo del fruto

glanduloso: que tiene órganos de secreción o glándulas

glaucescente: casi (volviéndose) glauco; verde grisáceo

glauco: cubierto con vello azulado o gris, como la cera que recubre a una ciruela

globoso: casi esférico (redondo)

lámina foliar: lámina foliar

marginado: con un margen completo, sin muescas

mate: sin brillo, opaco

mediano: en el medio, desarrollándose longitudinalmente

obtuso: romo, redondeado

pedicelo: el pedículo de la flor o el fruto

perianto: las partes vistosas de la flor: los pétalos (corola) o cáliz, o ambas juntas; *Galanthus* y *Sternbergia* (y muchas otras monocotiledóneas) no tienen cáliz, y la flor se define según los segmentos del perianto en vez de los pétalos

tubo del perianto: parte del perianto que se fusiona para formar un tubo, normalmente en la base

segmento del perianto: una parte del perianto; en *Galanthus* y *Sternbergia* hay seis segmentos para cada perianto (la flor)

pecíolo: el pedículo de la hoja

reniforme: en forma de riñón

revoluto: enrollado hacia el exterior desde el margen o el ápice

reflexo: encorbado abruptamente hacia abajo o hacia atrás

escapo: pedúnculo afilo surgiendo del suelo, con una o muchas flores

supervoluto (vernación): una hoja envuelve a la otra en la yema, o parcialmente de forma que las hojas se enrollan en la superficie adaxial en la yema

tubérculo: órgano de almacenamiento engrosado, subterráneo o parcialmente subterráneo

vernación: la disposición o arreglo de cada una de las hojas en la yema; en *Galanthus* es posible determinar el tipo de vernación después de la formación de la yema y durante la madurez, observando como se doblan las hojas en la base de la planta

ANNEX III: BIBLIOGRAPHY: PRIMARY REFERENCE SOURCES USED IN THE COMPILATION OF CHECKLISTS
For the genera:

Cyclamen, *Galanthus* and *Sternbergia*

ANNEXE III: BIBLIOGRAPHIE: PRINCIPALES SOURCES DE RÉFÉRENCE
Pour les genre:

Cyclamen, *Galanthus* et *Sternbergia*

ANEXO III: BIBLIOGRAFIA: PRINCIPALES FUENTES DE REFERENCIA
Para los géneros:

Cyclamen, *Galanthus* y *Sternbergia*

BIBLIOGRAPHY: PRIMARY REFERENCE SOURCES USED IN THE
COMPILATION OF CHECKLISTS

CYCLAMEN

Doorenbos, J. (1950). Taxonomy and nomenclature of *Cyclamen*. *Meded. Landbouwhogeschool* 50(2): 1–29.

Debussche, M. and Quézel, P. (1997). *Cyclamen repandum* Sibth. & Sm. en Petite Kabylie (Algérie): un témoin biogéographique méconnu au statut taxinomique incertain. *Acta Bot. Gallica* 144(1): 23–33.

Glasau, F. (1939). Monographie der Gattung *Cyclamen* auf morphologisch-cytologischer Grundlage. *Planta* 30: 507–550.

Grey-Wilson, C. (1988). *Cyclamen*. The Royal Botanic Gardens, Kew: in association with Christopher Helm and Timber Press.

Grey-Wilson, C. (1997). *The Genus Cyclamen*. London: B.T. Batsford Ltd.

Hildebrand, F.G.H. (1898). *Die Gattung Cyclamen* L. Jena: Gustav Fischer Verlag.

Hildebrand, F.G.H. (1907). Die *Cyclamen*-Arten als ein Beispiel für das Vorkommen nutzloser Verschiedenheiten im Pflanzenreich. *Beih. Bot. Centralbl.* 22: 143–196.

Meikle, R.D. and Sinnott, N.H. (1972). *Cyclamen*. In T.G. Tutin, et al. (eds.), *Flora Europaea* 3: 25–26. Cambridge: Cambridge University Press.

Meikle, R.D. (1978). *Cyclamen*. In P.H. Davis et al. (eds.) *Flora of Turkey and the East Aegean Islands* 6: 128–134. Edinburgh: University Press.

Meikle, R.D. (1985). *Cyclamen*. In *Flora of Cyprus* 2: 1077–1082. Royal Botanic Gardens, Kew: The Bentham-Moxon Trust.

Phitos, D., Strid, A., Snogerup, S. and Greuter, W. (1995). *The Red Data Book of Rare and Threatened Plants of Greece*. Athens: K. Michalas. World Wide Fund for Nature (WWF).

Popedimova, E.G. (1952). *Cyclamen*. In B.K. Shishkin and E.G. Bobrov. *Flora SSSR* 18: 279–290. Leningrad and Moscow: Izdatelstvo Akademii Nauk SSSR. Akademii Nauk SSSR, Botanicheskii Institut V.L. Komarova (Leningrad).

Raus, T. (1995). *Cyclamen persicum* Mill. In Phitos, D. et al. (eds.). *The Red Data Book of Rare and Threatened Plants of Greece*: 228–229. Athens: K. Michalas. World Wide Fund for Nature (WWF).

Saunders, D.E. (1959). *Cyclamen*, a gardener's guide to the genus. *Bull. Alpine Gard. Soc. Gr. Brit.* 27: 18–76.

Saunders, D.E. (1973 & 1975). Revised by R.D. Meikle and C. Grey-Wilson. *Cyclamen, the Genus in the Wild and in Cultivation.* The Alpine Garden Society, Woking, UK.

Schwarz, O. (1938). *Cyclamen* Studien. *Gartenflora* n.s., 1: 11–38.

Schwarz, O. (1955). Systematische Monographie der Gattung *Cyclamen* L. *Feddes Repert. Sp. Nov. Regni* Veg. 58: 234–283.

Schwarz, O. (1964). Systematische Monographie der Gattung *Cyclamen* L. *Feddes Repert. Sp. Nov. Regni* Veg. 69: 73–103.

GALANTHUS

Allen, J. (1891). Snowdrops. *J. Roy. Hort. Soc.* 13: 172–188.

Andriyenko, T.L., Melnik, V. I., Yakushina, L. A. (1992). Distribution and structure of *Galanthus nivalis* (Amaryllidaceae) coenopulations in Ukraine. *Bot. Zhurn. (Moscow & Leningrad)* 77(3): 101 - 107.

Artjushenko, Z.T. (1965). A contribution to the taxonomy of the genus *Galanthus. Bot. Zhurn. (Moscow & Leningrad)* 50(10): 1430–1447.

Artjushenko, Z.T. (1966). A critical review of the genus *Galanthus* L. *Bot. Zhurn. (Moscow & Leningrad)* 51(10): 1437–1451.

Artjushenko, Z.T. (1967). Taxonomy of the genus *Galanthus* L. *Daffodil Tulip Year Book* 32: 62–82, 87.

Artjushenko, Z.T. (1969). A critical review of the genus *Galanthus* L. *Plant Life* 25(2–4): 137–152.

Artjushenko, Z.T. (1970). *Amaryllidaceae J. St.- Hil. SSSR. Morphology, Systematics and Uses*: 41–83. Leningrad: Akademii Nauk SSSR. Botanicheskii Institut V.L.Komarova.

Artjushenko, Z.T. (1974). *Galanthus* L. (Amaryllidaceae) in Greece. *Ann. Mus. Goulandris* 2: 9–21.

Baker, J.G. (1888). *Handbook of the Amaryllideae*. London: George Bell & Sons.

Beck, G. von M. (1894). Die Schneeglocken, Eine monographische Skizze der Gattung *Galanthus*. Ill. *Gart.-Zeitung* 19: 45–58.

Boissier, E. (1882). *Flora Orientalis* 5: 144–146. Basel Geneve: H. Georg.

Bowles, E.A. (1918). Snowdrops. *J. Roy. Hort. Soc.* 43: 28–36.

Brickell, C.D. (1984). *Galanthus*. In P.H. Davis et al. (eds.), *Flora of Turkey and the East Aegean Islands* 8: 365–372. Edinburgh: University Press.

Brickell, C.D. (1986). *Galanthus* L. In S. M. Walters et al. (eds.), *The European Garden Flora* 1: 317–319. Cambridge: Cambridge University Press.

Budnikov, G. and Kricsfalusy, V. (1994). Bioecological study of *Galanthus nivalis* L. in the East Carpathians. *Thaiszia - J. Bot., Kosice* 4: 49–75.

Davis, A.P. (1997). Proposal to conserve the name *Galanthus elwesii* (Amaryllidaceae) with a conserved type. *Taxon* 46(3): 553–554.

Davis, A.P. (1999, in press). *The Genus Galanthus L.* The Royal Botanic Gardens, Kew: in association with Timber Press.

Davis, A.P., Mordak, H. and Jury, S.L. (1996). Taxonomic status of three Caucasian snowdrops: *Galanthus alpinus* Sosn., *G. bortkewitschianus* Koss and *G. caucasicus* (Baker) Grossh. Kew Bull 51(4): 741–752.

Delipavlov, D. (1971). The genus *Galanthus* L. (snowdrop) in Bulgaria. Izv. Bot. Inst. 21: 161–168.

Erik, S. and Demirkus, N. (1986). Contributions to the Flora of Turkey. *Doga Turk. J. Biol.* 10(1): 100–105.

Gottlieb-Tannenhain, P. von (1904). Studien über die Formen der Gattung *Galanthus*. *Abh. K. K. Zool.-Bot. Ges. Wien* 2(4): 1–93.

Grossheim, A.A. (1928). *Flora Kavkaza* 1: 288. Tiflis. Narodyni Komissariat Zemledeliya SSR Armenii Baku.

Grossheim, A.A. (1940). *Flora Kavkaza* edn. 2, 2: 193. Tiflis. Narodyni Komissariat Zemledeliya SSR Armenii Baku.

Halácsy, E. de. (1904). *Conspectus Florae Graecea* 3: 206. Leipzig: Wilhelm Engelmann.

Kamari, G. (1981). A biosystematic study of the genus *Galanthus* L. in Greece, part II (Cytology). *Bot. Chron.* 1(2): 60–98.

Kamari, G. (1982). A biosystematic study of the genus *Galanthus* L. in Greece, part I. *Bot. Jahrb. Syst.* 103(1): 107–135.

Kamari, G. (1995). *Galanthus ikariae* Baker. In Phitos, D. et al. (eds.). *The Red Data Book of Rare and Threatened Plants of Greece*: 290–291. Athens: K. Michalas. World Wide Fund for Nature (WWF).

Kemularia-Nathdaze, L.M. (1947). Galanthi generis species novae in "Flora Georgica" desccriptae. *Zametki Sist. Geogr. Rast.* 13: 6.

Koss, G.I. (1951). Species Caucasicae novae generis *Galanthus* L. *Bot. Mater. Gerb. Inst. Kom. a. Akad. Nauk. SSSR* 14: 130–138.

Lozina-Lozinskaya, A.S. (1935). *Galanthus*. In V.L. Komarov (ed.), *Flora SSSR* 4: 476–80. Leningrad: Izdatelstvo Akademii Nauk SSSR. Akademii Nauk SSSR Botanicheskii Institut V.L.Komarova.

Melnik, V.I. (1994). *Galanthus elwesii* Hook.f. (Amaryllidaceae) in the Ukraine. *Ukrajins'k. Bot. Zhurn.* 51(1): 29–33.

Papanicolaou, K. and Zacharof, E. (1983). Cytological notes and taxonomic comments on four *Galanthus* L. taxa from Greece. *Israel J. of Botany* 32: 23–32.

Stern, F.C. (1956). *Snowdrops and Snowflakes*. London: RHS.

Annex III: Bibliography - *Galanthus*

Traub, H.P. and Moldenke, H.N. (1948). The tribe Galantheae. *Herbertia* 14: 85–116.

Webb, D.A. (1978). The European species of *Galanthus* L. *Bot. J. Linn. Soc.* 76(4): 307–313.

Webb, D.A. (1980). *Galanthus*. In T.G. Tutin, et al. (eds.), *Flora Europaea* 5: 77-78. Cambridge: Cambridge University Press.

Wendelbo, P. (1970). Amaryllidaceae *Galanthus*. In K.L. Rechinger (cd.), *Flora Iranica*, No. 67: 6–7. Graz-Austria: Akademische Druck-u. Verlagsanstalt.

Zeybek, N. (1988). Taxonomic investigations on Turkish snowdrops. *Doga Tu. J. Botany* 12(1): 89–102.

Zeybek, N. and Sauer E. (1995). *Türkiye Kardelenleri (Galanthus L.) I /Beitrag Zur Türkischen Schneeglöckhen (Galanthus L.)* I. VSB Altinova - Karamürsel.

STERNBERGIA

Artjushenko, Z.T. (1970). *Amaryllidaceae J. St.- Hil. SSSR. Morphology, Systematics and Uses*: 83–97. Leningrad: Akademii Nauk SSSR. Botanicheskii Institut V.L.Komarova.

Baytop, T. and Mathew, B. (1984). *The Bulbous Plants of Turkey. An Illustrated Guide to the Bulbous Petaloid Monocotyledons of Turkey. Amaryllidaceae, Iridaceae, Liliaceae*. London: B.T. Batsford Ltd.

Feibrun, N. and Stearn, W.T. (1958). A revision of *Sternbergia* (Amaryllidaceae) in Palestine. *Bull. Research Council Israel, Sect. D: Botany* 6(3): 167–173.

Gorshkova, S.G. (1935). Sternbergia. In V.L. Komarov (ed.), *Flora SSSR* 4: 485–489. Leningrad: Izdatelstvo Akademii Nauk SSSR. Botanicheskii Institut V.L.Komarova.

Herbert, W. (1837). *Amaryllidaceae*: 186–188. London: James Ridgeway and Sons.

Kamari, G. and Artelari, R. (1990). Karyosystematic study of the genus *Sternbergia* (Amaryllidaceae) in Greece. I. South Aegean Islands. *Willdenowia* 19: 367–387.

Mathew, B. (1983). A review of the genus *Sternbergia*. *Plantsman* 5(1): 1–16.

Mathew, B. (1984). *Sternbergia*. In P.H. Davis et al. (eds.), *Flora of Turkey and the East Aegean Islands* 8: 360–364. Edinburgh: University Press.

Webb, D.A. (1980). *Sternbergia*. In T.G. Tutin, et al. (eds.), *Flora Europaea* 5: 76. Cambridge: Cambridge University Press.

Wendelbo, P. (1970). Amaryllidaceae - *Sternbergia*. In K.L. Rechinger (ed.), *Flora Iranica*, 67: 4–6. Graz-Austria: Akademische Druck-u. Verlagsanstalt.

Annex IV: List of extra synonyms for *Galanthus nivalis*
Annexe IV: Liste de synonyme supplementaire pour *Galanthus nivalis*
Anexo IV: Lista de otros sinonimos por *Galanthus nivalis*

ANNEX IV: LIST OF EXTRA SYNONYMS
Ordered alphabetically for:

Galanthus nivalis

ANNEXE IV: LISTE DE SYNONYMES SUPPLÉMENTAIRES
Par ordre alphabétique de tous les noms pour:

Galanthus nivalis

ANEXO IV: LISTA DE OTROS SINONIMOS
Presentados por orden alfabético para:

Galanthus nivalis

Annex IV: List of extra synonyms for *Galanthus nivalis*
Annexe IV: Liste de synonyme supplementaire pour *Galanthus nivalis*
Anexo IV: Lista de otros sinonimos por *Galanthus nivalis*

LIST OF EXTRA SYNONYMS FOR GALANTHUS NIVALIS

During the late nineteenth and early twentieth centuries, the use of standard conventions and guidelines for the naming of plants was neither so widely nor so strictly in force as it is today, particularly for names used in the horticultural literature. In many publications of this era there is no clear distinction made between botanical units of classification (e.g., species, subspecies and varieties), and cultivars (plants of garden/horticultural origin, e.g., resulting from hybridisation, or selection by man). This can be confusing, because in some cases it is unclear whether the author of the article was referring to a botanical unit of classification or a cultivar. This problem is compounded because these names often comply with the conditions in the ICBN (International Code of Botanical Nomenclature) for effective publication, which is retrospectively less rigorous when applied to earlier publications. In most circumstances, however, it is possible to deduce that the author was using the name without any desire to name a new species or variety, etc., but many authors did describe new taxa in the horticultural literature, such as in *The Garden* and *The Gardeners Chronicle*, and therefore these publications need to be checked carefully. Indeed, there are many examples of explicitly and legitimately published taxa, such as when the author described and discussed a plant of wild provenance, listed the characters that distinguish it from other taxa, or provided a Latin description. Other names are ambiguous, however, and it is often difficult or impossible to be clear as to the author's intentions. For the purposes of this work I have removed most of the equivocal names from the main body of this checklist. Those names that I think might be of garden origin have been included here. All these names represent synonyms of *G. nivalis*, and many of them have been reclassified as cultivars.

LISTE DE SYNONYMES SUPPLÉMENTAIRES POUR GALANTHUS NIVALIS

A la fin du 19ᵉ siècle et au début du 20ᵉ, l'utilisation de conventions et de lignes directrices pour nommer les nouvelles plantes n'était pas aussi courante et stricte que de nos jours, notamment dans la littérature horticole. De nombreuses publications de cette époque ne faisaient pas clairement la différence entre les entités de classification botanique (espèce, sous-espèce, variété, etc.) et les cultivars. C'est une source de confusion car dans certains cas, on ignore si l'auteur fait référence à une unité de classification botanique ou à un cultivar. Le problème est rendu encore plus complexe par le fait que ces noms suivent souvent les règles de publication du *International Code of Botanical Nomenclature*, lesquelles sont appliquées moins rigoureusement pour les publications plus anciennes. Dans la plupart des cas, on comprend que l'auteur n'a pas voulu nommer une nouvelle espèce ou variété; cependant, comme de nombreux auteurs ont effectivement décrit de nouveaux taxons dans la littérature horticole (*The Garden* et *The Gardeners Chronicle*, par exemple), ces publications doivent être soigneusement vérifiées. Enfait, de nombreux taxons sont publiés explicitement et légitimement, notamment lorsqu'un auteur décrit et détaille une plante d'origine sauvage, fait la liste des caractères la différenciant d'autres taxons, ou en donne une description en latin. Toutefois, certains noms sont ambigus et l'on ne peut que supputer les intentions de l'auteur. Dans le présent ouvrage, la plupart des noms les plus équivoques ont été supprimés de la liste principale. Les noms susceptibles d'être d'origine horticole y ont été inclus. Tous sont des synonymes de *G. nivalis*, souvent reclassés comme cultivars.

LISTA DE OTROS SINONIMOS DE GALANTHUS NIVALIS

A finales del Siglo XIX y principios del Siglo XX, el uso de prácticas y directrices normalizadas para la denominación de plantas no era tan corriente ni tan estricto como en la actualidad, en

Annex IV: List of extra synonyms for *Galanthus nivalis*
Annexe IV: Liste de synonyme supplementaire pour *Galanthus nivalis*
Anexo IV: Lista de otros sinonimos por *Galanthus nivalis*

particular para los nombres utilizados en los artículos sobre horticultura. En muchas publicaciones de esta época no se hace una distinción clara entre las unidades de clasificación botánica (p.e.: especies, subespecies y variedades) y los cultivares (plantas que se originaron en huertas o jardines, p.e.: resultantes de la hibridización o selección hechas por el hombre). Esto puede suscitar confusión porque en muchos casos no está claro si el autor del artículo se refiere a una unidad de clasificación botánica o a un cultivar. El problema se complica porque los nombres suelen cumplir las condiciones impuestas por el ICBN (Código Internacional de Nomenclatura Botánica) para la publicación, que retrospectivamente son menos rigurosas cuando se aplican a publicaciones más antiguas. En la mayoría de los casos, sin embargo, es posible deducir que el autor utilizaba el nombre sin deseo alguno de referirse a una nueva especie o variedad, etc., pero muchos autores sí describieron nuevos taxa en artículos sobre horticultura como los publicados en *The Garden* y *The Garden Chronicle*, y por consiguiente es necesario verificar esas publicaciones cuidadosamente. De hecho, hay muchos casos de taxaexplícita y legítimamente publicados, en los cuales el autor describe y examina una planta silvestre, enumera los caracteres que la distinguen de otros taxa, o hace una descripción en latín. Otros nombres, sin embargo, son ambiguos, y suele ser difícil o imposible tener una idea clara de las intenciones del autor. A los efectos de esta obra, se han eliminado la mayoría de los nombres equívocos de la lista. Se han incluido, en cambio, los nombres que podían tener un origen de jardín. Todos estos nombres son sinónimos de *G. nivalis* y muchos de ellos se han reclasificado como cultivares.

Galanthus nivalis L.

G. aestivalis Burb. in J. Roy. Hort. Soc. 13(2): 200 (1891). †
G. boydii Burb. in J. Roy. Hort. Soc. 13(2): 200 (1891). †
G. cathcartiae (hort. J.Allen) Burb. in J. Roy. Hort. Soc. 13(2): 200 (1891).
G. flavescens Boyd ex Ewbank in Garden (London) 39: 272 (1891). †
G. flavescens Burb. in Gard. Chron. ser. 3, 7: 268: (1890). †
G. flavescens Burb. in J. Roy. Hort. Soc. 13(2): 202 (1891). †
G. flavescens J.Allen in Garden (London) 40: 272 (1891). †
G. flavescens J.Allen in J. Roy. Hort. Soc. 13(2): 181 (1891).
G. lutescens hort. ex Correvon in Le Jardin 32: 139 (1888). †
G. lutescens J.Allen in Garden (London) 29: 75 (1886).
G. melvillei hort. ex Burb. in J. Roy. Hort. Soc. 13(2): 205, fig. 31 (1891). †
G. melvillei hort. ex Correvon in Le Jardin 32: 139 (1888). †
G. melvillei hort. ex Siebert & Voss in Vilm. Blumengärtenerei ed. 3, 1: 1006 (1895).†
G. melvillei [var.] *major* J.Allen in J. Roy. Hort. Soc. 13(2): 173 (1891). †
G. nivalis L. forma *virescens* Leichtlin in Gard. Chron. n.s., 11: 342 (1879).
G. nivalis L. [var.] *albus* J.Allen in J. Roy. Hort. Soc. 13(2): 182 (1891).
G. nivalis L. [var.] *cathcartiae* J.Allen in J. Roy. Hort. Soc. 13(2): 184 (1891).
G. nivalis L. var. *europaeus* Beck in Wiener Ill. Gart.-Zeitung 19: 50 (1894).
G. nivalis L. var. *europaeus* Beck forma *aestivalis* (Burb.) Beck in Wiener Ill. Gart.-Zeitung 19: 51 (1894).
G. nivalis L. var. *europaeus* Beck forma *albus* (J.Allen) Beck in Wiener Ill. Gart.-Zeitung 19: 50 (1894).
G. nivalis L. var. *europaeus* Beck forma *biflorus* Beck in Wiener Ill. Gart.-Zeitung 19: 52 (1894).
G. nivalis L. var. *europaeus* Beck forma *biscapus* Beck in Wiener Ill. Gart.-Zeitung 19: 52 (1894).

Annex IV: List of extra synonyms for *Galanthus nivalis*
Annexe IV: Liste de synonyme supplementaire pour *Galanthus nivalis*
Anexo IV: Lista de otros sinonimos por *Galanthus nivalis*

G. nivalis L. var. *europaeus* Beck forma *candidus* Beck in Wiener Ill. Gart.-Zeitung 19: 52 (1894).

G. nivalis L. var. *europaeus* Beck forma *cathcartiae* (J.Allen) Beck in Wiener Ill. Gart.-Zeitung 19: 52 (1894).

G. nivalis L. var. *europaeus* Beck forma *pallidus* (Smith) Beck in Wiener Ill. Gart.-Zeitung 19: 51 (1894).

G. nivalis L. var. *europaeus* Beck forma *platytepalus* Beck in Wiener Ill. Gart.-Zeitung 19: 50, fig. 1,2 (1894).

G. nivalis L. var. *europaeus* Beck forma *plenissimus* Beck in Wiener Ill. Gart.-Zeitung 19: 50 (1894).

G. nivalis L. var. *europaeus* Beck forma *poculiformis* (hort.) Beck in Wiener Ill. Gart.-Zeitung 19: 50 (1894). †

G. nivalis L. var. *europaeus* Beck forma *sandersii* (Harpur-Crewe) Beck in Wiener Ill. Gart.-Zeitung 19: 50 (1894).

G. nivalis L. var. *europaeus* Beck forma *stenotepalus* Beck in Wiener Ill. Gart.-Zeitung 19: 50, fig. 1,1 (1894).

G. nivalis L. var. *europaeus* Beck forma *trifolius* Beck in Wiener Ill. Gart.-Zeitung 19: 52 (1894).

G. nivalis L. var. *europaeus* Beck forma *virescens* (Leichtlin) Beck in Wiener Ill. Gart.-Zeitung 19: 51, fig. 1,4 (1894).

G. nivalis L. var. *europaeus* Beck forma *viridans* Beck in Wiener Ill. Gart.-Zeitung 19: 51, fig. 1,3 (1894).

G. nivalis L. var. *lutescens* hort. ex Baker, Handb. Amaryll.: 17 (1888). †

G. nivalis L. var. *lutescens* Mallett in Garden (London) 67: 87 (1905). †

G. nivalis L. [var.] *melvillei* Bowles in J. Roy. Hort. Soc. 43: 30, fig. 2 (1918). †

G. nivalis L. var. *melvillei* Harpur-Crewe in Gard. Chron. n.s., 11: 237 (1879).

G. nivalis L. var. *poculiformis* (Beck) Bowles in J. Roy. Hort. Soc. 43: 30 (1918). †

G. nivalis L. var. *poculiformis* hort. ex Baker, Handb. Amaryll.: 17 (1888). †

G. nivalis L. var. *poculiformis* Mallett in Garden (London) 67: 87 (1905). †

G. nivalis L. var. *sandersii* Harpur-Crewe in Gard. Chron. n.s., 11: 342 (1879).

G. nivalis L. [var.] *serotinus* hort. ex Correvon in Le Jardin 32: 140 (1888).

G. pallidus Smith [?] ex Burb. in J. Roy. Hort. Soc. 13(2): 206 (1891). †

G. poculiformis hort. ex Burb. in J. Roy. Hort. Soc. 13(2): 207 (1891). †

G. poculiformis hort. ex Correvon in Le Jardin 32: 139 (1888). †

G. poculiformis J.Allen in Garden (London) 29: 75 (1886). †

G. serotinus D.Melville in Gard. Chron. n.s., 15: 181 (1881). †

G. serotinus hort. [Dunrobin] ex. Burb. in J. Roy. Hort. Soc. 13(2): 209 (1891). †

G. umbrensis [sic] hort. [Dammann] ex Burb. in J. Roy. Hort.Soc. 13(2): 209 (1891).†

G. umbricus J.Allen in J. Roy. Hort. Soc. 13(2): 184 (1891). †

G. umbricus Wolley Dod in Gard. Chron. ser. 3, 7: 207 (1890). †

G. virescens Anon in Garden (London) 25: 371 (1884).

G. virescens Burb. J. Roy. Hort.Soc. 13(2): 209, fig 33. (1891). †

G. virescens Ewbank in Garden (London) 39: 273, fig. on p. 276 (1891). †

G. virescens J.Allen in Garden (London) 29: 75 (1886). †

G. virescens J.Allen in Garden (London) 40: 272 (1891). †

G. virescens (Leichtlin) Correvon in Le Jardin 2: 140 (1888).

G. warei Burb. in J. Roy. Hort. Soc. 13(2): 210 (1891). †

Annex IV: List of extra synonyms for *Galanthus nivalis*
Annexe IV: Liste de synonyme supplementaire pour *Galanthus nivalis*
Anexo IV: Lista de otros sinonimos por *Galanthus nivalis*

G. warei J.Allen in J. Roy. Hort. Soc. 13(2): 183 (1891[March 10th]); J.Allen in Garden (London) 40: 273 (1891[September 19th]).

† Names not validly published or names invalid
† Noms non publiés validement ou noms non valides
† Nombres publicados que carecen de validez o nombres inválidos

Annex V: Summary of accepted names
Annexe V: Résumé de noms acceptés
Anexo V: Resumen de los nombres aceptados

ANNEX V: SUMMARY OF ACCEPTED NAMES
Alphabetically arranged for the genera:

Cyclamen, *Galanthus* and *Sternbergia*

ANNEXE V: RÉSUMÉ DES NOMS ACCEPTÉS
Par ordre alphabétique de tous les noms pour les genres

Cyclamen, *Galanthus* et *Sternbergia*

ANEXO V: RESUMEN DE LOS NOMBRES ACEPTADOS
Presentados por orden alfabético para los géneros:

Cyclamen, *Galanthus* y *Sternbergia*

Annex V: Summary of accepted names - *Cyclamen*
Annexe V: Résumé des noms acceptés - *Cyclamen*
Anexo V: Resumen de los nombres aceptados - *Cyclamen*

CYCLAMEN BINOMIALS IN CURRENT USE

CYCLAMEN BINOMES ACTUELLEMENT EN USAGE

CYCLAMEN BINOMIALES UTILIZADOS NORMALMENTE

1.　**Cyclamen africanum** Boiss. & Reut.
2.　**Cyclamen balearicum** Willk.
3.　**Cyclamen cilicium** Boiss. & Heldr.
3.1.　**Cyclamen cilicium** Boiss. & Heldr. forma **album** E.Frank & Koenen
3.2.　**Cyclamen cilicium** Boiss. & Heldr. forma **cilicium**
4.　**Cyclamen colchicum** (Albov) Albov
5.　**Cyclamen coum** Mill.
5.1.　**Cyclamen coum** Mill. subsp. **caucasicum** (K.Koch) O.Schwarz
5.2.　**Cyclamen coum** Mill. subsp. **coum** forma **albissimum** R.H.Bailey, Koenen, Lillywh. & P.J.M.Moore
5.3.　**Cyclamen coum** Mill. subsp. **coum** forma **coum**
5.4.　**Cyclamen coum** Mill. subsp. **coum** forma **pallidum** Grey-Wilson
5.5.　**Cyclamen coum** Mill. subsp. **elegans** (Boiss. & Buhse) Grey-Wilson
6.　**Cyclamen creticum** (Dörfl.) Hildebr.
6.1.　**Cyclamen creticum** (Dörfl.) Hildebr. forma **creticum**
6.2.　**Cyclamen creticum** (Dörfl.) Hildebr. forma **pallide-roseum** Grey-Wilson
7.　**Cyclamen cyprium** Kotschy
8.　**Cyclamen graecum** Link
8.1.　**Cyclamen graecum** Link subsp. **anatolicum** Ietsw.
8.2.　**Cyclamen graecum** Link subsp. **graecum** forma **album** R.Frank
8.3.　**Cyclamen graecum** Link subsp. **graecum** forma **graecum**
8.4.　**Cyclamen graecum** Link subsp. **mindleri** (Heldr.) A.P.Davis & Govaerts
9.　**Cyclamen hederifolium** Aiton
9.1.　**Cyclamen hederifolium** Aiton var. **confusum** Grey-Wilson
9.2.　**Cyclamen hederifolium** Aiton var. **hederifolium** forma **albiflorum** (Jord.) Grey-Wilson
9.3.　**Cyclamen hederifolium** Aiton var. **hederifolium** forma **hederifolium**
10.　**Cyclamen intaminatum** (Meikle) Grey-Wilson
11.　**Cyclamen libanoticum** Hildebr.
12.　**Cyclamen mirabile** Hildebr.
12.1.　**Cyclamen mirabile** Hildebr. forma **mirabile**
12.2.　**Cyclamen mirabile** Hildebr. forma **niveum** J.White & Grey-Wilson
13.　**Cyclamen parviflorum** Poped.
13.1.　**Cyclamen parviflorum** Poped. var. **parviflorum**
13.2.　**Cyclamen parviflorum** Poped. var. **subalpinum** Grey-Wilson
14.　**Cyclamen persicum** Mill.
14.1.　**Cyclamen persicum** Mill. var. **autumnale** Grey-Wilson
14.2.　**Cyclamen persicum** Mill. var. **persicum** forma **albidum** (Jord.) Grey-Wilson
14.3.　**Cyclamen persicum** Mill. var. **persicum** forma **persicum**
14.4.　**Cyclamen persicum** Mill. var. **persicum** forma **puniceum** Grey-Wilson
14.5　**Cyclamen persicum** Mill. var. **persicum** forma **roseum** Grey-Wilson, nom. provis.
15.　**Cyclamen pseudibericum** Hildebr.
15.1.　**Cyclamen pseudibericum** Hildebr. forma **pseudibericum**
15.2.　**Cyclamen psuedibericum** Hildebr. forma **roseum** Grey-Wilson

Annex V: Summary of accepted names - *Cyclamen*
Annexe V: Résumé de noms acceptés - *Cyclamen*
Anexo V: Resumen de los nombres aceptados - *Cyclamen*

16. **Cyclamen purpurascens** Mill.
16.1. **Cyclamen purpurascens** Mill. forma **album** Grey-Wilson
16.2. **Cyclamen purpurascens** Mill. forma **purpurascens**
17. **Cyclamen repandum** Sm.
17.1. **Cyclamen repandum** Sm. subsp. **peloponnesiacum** var. **peloponnesiacum** (Grey-Wilson) Grey-Wilson
17.2. **Cyclamen repandum** Sm. subsp. **peloponnesiacum** Grey-Wilson var. **vividum** (Grey-Wilson) Grey-Wilson
17.3. **Cyclamen repandum** Sm. subsp. **repandum** var. **baborense** Debussche & Quézel
17.4 **Cyclamen repandum** Sm. subsp. **repandum** var. **repandum** forma **album** Grey-Wilson
17.5. **Cyclamen repandum** Sm. subsp. **repandum** var. **repandum** forma **repandum**
17.6. **Cyclamen repandum** Sm. subsp. **rhodense** (Miekle) Grey-Wilson
18. **Cyclamen rohlfsianum** Asch.
19. **Cyclamen somalense** Thulin & Warfa
20. **Cyclamen trochopteranthum** O.Schwarz
20.1. **Cyclamen trochopteranthum** O.Schwarz forma **leucanthum** Grey-Wilson
20.2. **Cyclamen trochopteranthum** O.Schwarz forma **trochopteranthum**

Hybrids

1. **Cyclamen** × **atkinsii** T.Moore
2. **Cyclamen** × **drydeniae** Grey-Wilson
3. **Cyclamen** × **hildebrandii** O.Schwarz
4. **Cyclamen** × **meiklei** Grey-Wilson
5. **Cyclamen** × **saundersiae** Grey-Wilson
6. **Cyclamen** × **schwarzii** Grey-Wilson
7. **Cyclamen** × **wellensiekii** Ietsw.
8. **Cyclamen** × **whiteae** Grey-Wilson

Annex V: Summary of accepted names - *Galanthus*
Annexe V: Résumé des noms acceptés - *Galanthus*
Anexo V: Resumen de los nombres aceptados - *Galanthus*

GALANTHUS BINOMIALS IN CURRENT USE

GALANTHUS BINOMES ACTUELLEMENT EN USAGE

GALANTHUS BINOMIALES UTILIZADOS NORMALMENTE

1.   **Galanthus alpinus** Sosn.
1.1. **Galanthus alpinus** Sosn. var. **alpinus**
1.2. **Galanthus alpinus** Sosn. var. **bortkewitschianus** (Koss) A.P.Davis
2.   **Galanthus angustifolius** Koss
3.   **Galanthus cilicicus** Baker
4.   **Galanthus elwesii** Hook.f.
5.   **Galanthus fosteri** Baker
6.   **Galanthus gracilis** Čelak.
7.   **Galanthus ikariae** Baker
8.   **Galanthus koenenianus** Lobin, C.D.Brickell & A.P.Davis
9.   **Galanthus krasnovii** A.P.Khokhr.
10.   **Galanthus lagodechianus** Kem.-Nath.
11.   **Galanthus nivalis** L.
12.   **Galanthus peshmenii** A.P.Davis & C.D.Brickell
13.   **Galanthus platyphyllus** Traub & Moldenke
14.   **Galanthus plicatus** M.Bieb.
14.1. **Galanthus plicatus** M.Bieb. subsp. **byzantinus** (Baker) D.A.Webb
14.2. **Galanthus plicatus** M.Bieb. subsp. **plicatus**
15.   **Galanthus reginae-olgae** Orph.
15.1. **Galanthus reginae-olgae** Orph. subsp. **reginae-olgae**
15.2. **Galanthus reginae-olgae** Orph. subsp. **vernalis** Kamari
16.   **Galanthus rizehensis** Stern
17.   **Galanthus transcaucasicus** Fomin
18.   **Galanthus woronowii** Losinsk.

Hybrids

1.   **Galanthus** × **allenii** Baker
2.   **Galanthus** × **grandiflorus** Baker

Cultivars

1.   **Galanthus nivalis** L. **'Flore Pleno'**

Annex V: Summary of accepted names - *Sternbergia*
Annexe V: Résumé de noms acceptés - *Sternbergia*
Anexo V: Resumen de los nombres aceptados - *Sternbergia*

STERNBERGIA BINOMIALS IN CURRENT USE

STERNBERGIA BINOMES ACTUELLEMENT EN USAGE

STERNBERGIA BINOMIALES UTILIZADOS NORMALMENTE

1. **Sternbergia candida** B.Mathew & T.Baytop
2. **Sternbergia clusiana** (Ker Gawl.) Ker Gawl. ex Spreng.
3. **Sternbergia colchiciflora** Waldst. & Kit.
4. **Sternbergia fischeriana** (Herb.) M.Roem.
5. **Sternbergia greuteriana** Kamari & R.Artelari
6. **Sternbergia lutea** (L.) Ker Gawl. ex Spreng.
7. **Sternbergia pulchella** Boiss. & Blanche
8. **Sternbergia schubertii** Schenk
9. **Sternbergia sicula** Tineo ex Guss.

Annex VI: A new combination in Cyclamen
Annexe VI: Une nouvelle combinaison dans Cyclamen
Anexo VI: Una nueva combinación en Cyclamen

ANNEX VI: A NEW COMBINATION IN CYCLAMEN

ANNEXE VI: UNE NOUVELLE COMBINAISON DANS CYCLAMEN

ANEXO VI: UNA NUEVA COMBINACIÓN EN CYCLAMEN

Annex VI: A new combination in Cyclamen
Annexe VI: Une nouvelle combinaison dans Cyclamen
Anexo VI: Una nueva combinación en Cyclamen

A NEW COMBINATION IN CYCLAMEN
A.P.Davis and R.Govaerts

During the compilation of this checklist it became evident that the name *Cyclamen graecum* Link subsp. *candicum* Ietsw. needs to be replaced because an earlier name exists for this taxon. Under the rules of the International Code of Botanical Nomenclature (ICBN) the earliest name of any taxon below the rank genus and at the same rank (e.g. species, subspecies or variety) has priority. The taxon *Cyclamen persicum* Mill. subsp. *mindleri* (Heldr.) Knuth. (published in 1905) has priority over *Cyclamen graecum* Link subsp. *candicum* Ietsw. (published in 1990). Grey-Wilson (1997, pages 101, 105, 172) clearly considers that *Cyclamen mindleri* Heldr. (*sic* - published as *Cyclaminos mindleri* Heldr.), *Cyclamen persicum* Mill. subsp. *mindleri* (Heldr.) Knuth, and *Cyclamen graecum* Link subsp. *candicum* are the same taxon. It is therefore necessary to make the new combination of *Cyclamen graecum* Link subsp. *mindleri* to replace the illegitimate *Cyclamen graecum* subsp. *candicum.*

UNE NOUVELLE COMBINAISON DANS CYCLAMEN
A.P.Davis and R.Govaerts

Au cours de la compilation de la présente Liste, il est clairement apparu que le nom *Cyclamen graecum* Link subsp. *candicum* Ietsw. doit être remplacé, un nom antérieur existant pour ce taxon. Selon les règles du Code international de nomenclature botanique, le nom le plus ancien de tout taxon infragénérique de même rang (par exemple espèce, sous-espèce ou variété) a la priorité. Le taxon *Cyclamen persicum* Mill. subsp. *mindleri* (Heldr.) Knuth. (publié en 1905) a la priorité sur *Cyclamen graecum* Link subsp. *candicum* Ietsw. (publié en 1990). Grey-Wilson (1997, pages 101, 105, 172) considère clairement que *Cyclamen mindleri* Heldr. (sic - publié comme *Cyclaminos mindleri* Heldr.), *Cyclamen persicum* Mill. subsp. *mindleri* (Heldr.) Knuth, et *Cyclamen graecum* Link subsp. *candicum* sont le même taxon. Il faut donc établir la nouvelle combinaison *Cyclamen graecum* Link subsp. *mindleri* pour remplacer *Cyclamen graecum* subsp. *candicum*, qui n'est pas valable.

UNA NUEVA COMBINACIÓN EN CYCLAMEN
A.P.Davis and R.Govaerts

Durante la compilación de esta Lista se llegó a la conclusión de que el nombre *Cyclamen graecum* Link subsp. *candicum* Ietsw. debía remplazarse ya que este taxón se conocía por otro nombre. En virtud del Reglamento del Código Internacional de Nomenclatura Botánica el nombre más antiguo de cualquier taxón por debajo del nivel de género y al mismo nivel (por ejemplo, especie, subespecie o variedad) tiene prioridad. El taxón *Cyclamen persicum* Mill. subsp. *mindleri* (Heldr.) Knuth. (publicado en 1905) tiene prioridad sobre *Cyclamen graecum* Link subsp. *candicum* Ietsw. (publicado en 1990). Grey-Wilson (1997, páginas 101, 105, 172) estima que *Cyclamen mindleri* Heldr. (sic – publicado como *Cyclaminos mindleri* Heldr.), *Cyclamen persicum* Mill. subsp. *mindleri* (Heldr.) Knuth, y *Cyclamen graecum* Link subsp. *candicum* son el mismo taxón.

Annex VI: A new combination in Cyclamen
Annexe VI: Une nouvelle combinaison dans Cyclamen
Anexo VI: Una nueva combinación en Cyclamen

En consecuencia, es preciso establecer una nueva combinación de *Cyclamen graecum* Link subsp. *mindleri* para remplazar la denominación ilegítima de *Cyclamen graecum* subsp. *candicum*.

Cyclamen graecum Link subsp. **mindleri** (Heldr.) A.P.Davis & Govaerts, **comb. nov.**

Basionym: *Cyclaminos mindleri* Heldr. in Bull. Herb. Boiss. 6: 386 (1898).
                Cyclamen persicum Mill. subsp. *mindleri* (Heldr.) Knuth, Pflanzenr. 4, 237: 248 (1905).

Synonyms: *Cyclamen pseudograecum* Hildebr. in Gartenflora 60: 629 (1911).
                Cyclamen graecum Link subsp. *candicum* Ietsw. in Cyclamen Soc. J. 14(2): 50 (1990).

Notes

Notes

Notes